Hans-C. Flemming, Gill G. Geesey (Eds.)

Biofouling and Biocorrosion in Industrial Water Systems

Proceedings of the International Workshop on Industrial Biofouling and Biocorrosion, Stuttgart, September 13-14, 1990

Springer-Verlag
Berlin Heidelberg NewYork
London Paris Tokyo
Hong Kong Barcelona Budapest

Dr. Hans-Curt Flemming
Universität Stuttgart
Institut für Siedlungswasserbau
Chemische Abteilung
Bandtäle 1
7000 Stuttgart 80
FRG

Dr. Gill Gregory Geesey
Dept. of Microbiology
California State University
Long Beach
CA 90840
USA

ISBN-13 : 978-3-642-76545-2   e-ISBN-13 : 978-3-642-76543-8
DOI: 10.1007/978-3-642-76543-8

ACKNOWLEDGEMENT

The Workshop on Biofouling and Biocorrosion in Industrial Water Systems has been financially supported by the Willy-Hager-Stiftung, Stuttgart, which is gratefully acknowledged.

PREFACE

The Workshop on Biofouling and Biocorrosion in Stuttgart, Germany in September 1990 as well as previous workshops on Biofouling of Industrial Water Systems in the United States have been offered to bring together individuals from industrial, academic, and government sectors. They provide a good opportunity to discuss problems related to microbial fouling and corrosion and identify various approaches that have been used to obtain satisfactory solutions to specific problems in the past. The exchange of information between participants and speakers that occurred at these workshops resulted in an improved level of understanding of the problems. That should lead to improved protection of equipment that has become fouled or corroded through surface-associated microbial activities.

Biofouling and biocorrosion represent a significant expense to operators of industrial water systems throughout the world. The recently completed pipeline that transports crude oil from Prudhoe Bay, Alaska across hundreds of miles of sensitive terrestrial habitats to the port of Valdez already suffers from microbially-influenced corrosion (MIC). Billions of dollars are needed to repair or replace damaged sections of pipe. Fouling and corrosion of ship hulls leads to increased fuel costs because of increased drag resistance. Biofouling and biocorrosion of heat exchangers and water circulation systems in the power utilities industry are responsible for reduced performance and operating efficiency. In some cases, these problems result in safety violations and in unacceptable environmental risks and consequences.

In spite of the widespread occurrence and serious nature of these microbially-mediated problems, relatively little attention and resources are directed to obtain an understanding of the fundamental reactions that are involved. More often, efforts and resources are applied to obtain "quick-fix" solutions to the problems which provide little or no advancement of understanding the underlying processes involved.

This text follows the tradition of the previously-published text "Biofouling of Industrial Water Systems: A Problem-Solving Approach" (1) to make available information that was presented at the above mentioned workshop. It was prepared as an aid to operators of industrial water systems in understanding some of the important surface-associated microbiological problems that they are likely to encounter. The text presents the state-of-knowledge of microbial growth in biofilms as it relates to fouling and corrosion in various industrial water systems. Phenomena that influence microbial growth and activities on conduits that transport reclaimed, potable and high-purity water are described, including the problems of fouling of reverse osmosis membranes.

We hope that readers will be encouraged to apply the knowledge obtained from this and other sources to approach microbial fouling and corrosion problems that they encounter. We also hope that they will participate in future workshops and share their experiences with the growing number of workers in that field.

Long Beach and Stuttgart, October 1990

Gill G. Geesey
Hans-Curt Flemming

(1)   Mittelman MW, Geesey GG (eds)(1987) Biological fouling of industrial water systems: A problem solving approach. Water Micro Associates, PO Box 28848 San Diego, CA 92128-0848

CONTENTS

INTRODUCTION:
BIOFILMS AS A PARTICULAR FORM OF MICROBIAL LIFE

Hans-Curt Flemming

Institut für Siedlungswasserbau, Wassergüte und Abfallwirtschaft
der Universität Stuttgart
Bandtäle 1, D-7000 Stuttgart 80

Both biofouling and biocorrosion are phenomena which are linked to the existence, properties and activity of biofilms. If we want to control and prevent these phenomena, we have to understand the dynamics of *biofilms*.

Biofilms are the result of the adhesion and growth of microorganisms at interfaces. The term "biofilm" refers to a microbial form of life which can vary in the extreme: different microorganisms attach to different surfaces in different environments. The range of biofilm existence is large, as are the definitions of what a biofilm is.

Biofilms are part of our daily life. When brushing our teeth, we intend nothing else but to remove the biofilm which could cause caries. Biofilms are ubiquitous. There is almost no surface which are is colonized or cannot be colonized under suitable conditions. We still do not know any material or coating which can withstand microbial colonization on a long term basis. Biofilms are a very successful form of the organization of microbial life, and are capable of adapting to extreme environmental conditions (Table 1):

Table 1 Some examples for the range of microbial existence in biofilms

---

Temperature
from -12°C cold, saline environment (1)
to >110°C sulfate reducers in hydrothermal vents (2)

pH range
from 0 *Thiobacillus ferrooxidans* (1)
to >13 *Plectonema nostocorum*, "natron bacteria" (3)

Hydrostatic pressure
from 0 (various bacteria)
to > 1400 bar "barophilic bacteria" (1)

Redox potential
Total range of the redox stability of water
Biofilms on electrodes

Salinity
from 0 (bacteria in distilled water) (4)
to saturated saline solutions (5)

Nutrients
from 10 μg/l ultrapure water systems (6)
to life directly on nutrient sources "biodegradation"

H.-C. Flemming · G. G. Geesey (Eds.)
Biofouling and Biocorrosion in Industrial Water Systems
Proceedings of the International Workshop on
Industrial Biofouling and Biocorrosion, Stuttgart, Sept. 13-14, 1990
© Springer-Verlag Berlin Heidelberg 1991

Table 1, continued

Surface materials
    Metals including "toxic" metals such as copper and silver;
    Gallium arsenide, concrete, plastics, glass, minerals, oils,
    plants, animal tissues, bones etc.
Irradiation
    Biofilms on UV lamps in water   (7)
    Biofilms on radioactive materials   (8, 9)
Biocide concentrations
    > 2 mg free chlorine   (10)
    Biofilms in disinfection lines   (11)
Dryness
    "osmophiles", e.g. filamentous fungi   (12)

Biofilms play an important role in nature. Their composition depends on the environmental conditions. The first records of life on earth have been found in the form of petrified biofilms in stromatolithic rocks (13). Stromatolithes arise from the mineralization of biofilms - microbial mats - which have been growing one on top of each other for millions of years. The oldest ones date back 3.5 billion years (14, 15).

Costerton et al. (16) estimate that more than 90% of the microorganisms in the biosphere live in biofilms. In the sessile mode of existence microbes are offered certain advantages to survival not available to free-living ("planctonic") microorganisms (Table 2):

Table 2 Ecological advantages of the biofilm mode of growth for microorganisms

---

*   Scavenging and enrichment of nutrients in the gel matrix
*   Protection from
    -     short term pH-fluctuations, salt- and biocide concentration shocks
    -     shear forces cells are kept in place and can scavenge nutrients from the flowing medium
    -     dehydration
*   Pool of genetical information
*   Development of microconsortia
    -     symbiosis e.g. in lichens, stromatolithes, etc.
    -     utilization of less readily degradable substrates by specialized organisms e.g. degradation of cellulose or xenobiotics; nitrification
    -     creation of ecological niches e.g. anaerobic zones under aerobic biofilms in aerobic environments; nutrient-rich compartments in oligotrophic systems
    -     feasibility of gene transfer because of the long retention times of the microorganisms

---

Hydrophobic biofilms that occur at the air-water-interface ("neuston layer") are of a great ecological significance (17) because of their influence on the exchange of matter between the gas and the water phases.

Much of the degradation of organic matter in soil and aquatic habitats is mediated by microbes in biofilms. Sewage water treatment technology throughout the world depends on biofilms for efficient removal of organic matter from waste water.

An early example of biofilm technology is the production of an important gunpowder component: Friedrich II of Prussia ordered the construction of lime walls over which liquid manure was poured. Nitrifying biofilms inside the stones converted the ammonium to saltpetre ("sal petrae" = salt of the stones) - calcium nitrate - which was boiled with ashes to obtain potassium nitrate (18).

The cases considered in this book concentrate on biofilms established at a solid-water interface. Biofilms consist mainly of:

*   water (up to 95% of the total biofilm mass)
*   extracellular polymeric substances EPS (up to 95% of the dry biofilm mass)
*   microorganisms
*   embedded particles
*   dissolved substances.

The role of the extracellular polymeric substances (EPS) is very significant (19) and their rheological properties influence the physical properties of biofilms (20). Biofilms can be considered as EPS-matrices which include immobilized microorganisms, in well organized microconsortia of different species capable of participating in complex interactions (21). Biofilms introduce a gel phase between the solid surface and the bulk water phase. This can change the hydrodynamic properties of a system dramatically. If biofilms grow on heat exchanger surfaces, they may drastically decrease the heat transfer rate because they shield the metal surface from tangential flow and thus lead to a decreased convection of water in the proximity of the wall (22). In pipelines, biofilms lead to an increased fluid friction resistance (e.g. 23), and the same effect occurs when bio-films grow on ship hulls (20). In water treatment systems (see chapters 4 and 7) biofilms may contaminate the water, colonize ion exchangers (24), and block separation membranes (25; see chapter 6). They may attack their support, thus leading to microbially induced corrosion of metals MIC, "biocorrosion" (see chapters 8 - 11). Biofilms are responsible for a significant amount of weathering of minerals (26) and thus for the biodeterioration of building materials (27, 28). In sewers, sulfate reducing bacteria from submerged biofilms produce H_2S, that is metabolized by sulfur oxidizing bacteria which live at the concrete-air-interface. The resulting sulfuric acid causes damage in the sewer systems (29) costing enormous sums of money. In porous media e.g. soils, aquifers, filters they can cause clogging (30).

In medicine, the possibility of microbial colonization of devices and implants represents a severe problem (31) and requires special precautions. Pathogenic bacteria such as *Legionella* (32, 33) and Mycobacteria (34) are frequently found in biofilms; the latter are common members of biofilms on metal mouthpieces of musical instruments ("trumpet bacteria") and telephones (35).

4

The effects of biofilms indicate their significance in natural and industrial processes. The following chapters will introduce the reader to biofilm-related problems and offer problem solving approaches that can be applied in industrial water systems.

REFERENCES

1   Atlas RM, Bartha R (1987) Microbial ecology: fundamentals and applications. Benjamin/Cummings, Menlo Park

2   Stetter KO, Segerer A, Zillig W, Huber G, Fiala G, Huber R, König H (1985) Extremely thermophilic sulfur-metabolizing archebacteria. System. Appl. Microbiol. 7, 393-397

3   Tindall BJ (1986) The natronbacteria: alkaliphilic members of the aerobic, halophilic archebacteria. In: Kander OU, Zillig W (eds) Archebacteria '85. G. Fischer Verlag, Stuttgart; 410

4   Favero MS, Carson LA, Bond WW, Petersen NJ (1971) *Pseudomonas aeruginosa*: growth in distilled waters from hospitals. Science 173, 836-838

5   Oren A (1986) Relationships of extremely halophilic bacteria towards divalent cations. In: Megusar F, Gantar M (eds) Microbial Ecology. Proc. 4th Int. Symp. on Microb. Ecol., 24-29 Aug. 1986, Ljubljana, 52-58

6   Morita RY (1985) Starvation and miniaturisation of heterotrophs with special emphasis on maintenance of the starved viable state. In: Fletcher M, Floodgate G (eds) Bacteria in their natural environments: the effect of nutrient conditions. Soc. Gen. Microbiol., U.K.; 111-130

7   Kreft P, Scheible OK, Venosa A (1986) Hydraulic studies and cleaning evaluations of ultraviolet disinfection units. J. Water Poll. Contr. Fed. 58, 1129-1137

8   Murray RGE (1986) The biology and ecology of the Deinococcae. In: Megusar F, Gantar M (eds) Microbial Ecology. Proc. 4th Int. Symp. on Microb. Ecol., 24-29 Aug. 1986, Ljubljana, 153-158

9   Lessel T, Mötsch H, Hennig E (1975) Experience with a pilot plant for the irradiation of sewage sludge. In: Radiation for a clean environment. Int. Atom. En. Ag. Vienna, 1975; 447-463

10   Characklis WG (1990) Microbial biofouling control. In: Characklis WG, Marshall KC (eds) Biofilms. John Wiley, New York, 1990; 585-634

11   Exner M, Tuschewitzki GJ, Thofern E (1983) Untersuchungen zur Wandbesiedlung der Kupferrohrleitung einer zentralen Desinfektionsmitteldosieranlage. Zbl. Bakt. Hyg. B 177, 170-181

12   Jennings DH (1990) Osmophiles. In: Edwards C (ed) Microbiology of extreme environments. McGraw-Hill, London, 117-146

13   Ferris FG, Shotyk W, Fyfe WS (1989) Mineral formation and decomposition by microorganisms. In: Beveridge TJ, Doyle RJ (eds) Metal ions and bacteria. John Wiley, New York, 413-441

14   Schopf JW, Hayes JM, Walter MR (1983) Evolution on earth's earliest ecosystems: recent progress and unsolved problems. In: Schopf JW (ed) Earth's earliest biosphere, Princeton Univ. Press, New Jersey; 361-384

15   Knoll AH (1989) The paleomicrobiological information in proterozoic rocks. In: Cohen Y, Rosenberg E (eds) Microbial mats. Am. Soc. Microbiol., Washington; 469-484

16   Costerton JW, Cheng KJ, Geesey GG, Ladd TJ, Nickel JC, Dasgupta M, Marrie T (1987) Bacterial biofilms in nature and disease. Ann. Rev. Microbiol. 41, 435-464

17   Kjelleberg S (1985) Mechanisms of bacterial adhesion at gas-liquid interfaces. In: Savage CD, Fletcher MM (eds) Bacterial adhesion. Plenum Press, New York; 163-194

18   Schlegel HG (1985) Allgemeine Mikrobiologie. Thieme, Stuttgart

19   Geesey GG (1982) Microbial exopolymers: Ecological and economic considerations. ASM News 48, 9-14

20   Christensen BE, Characklis WG (1990) Physical and chemical properties of biofilms. In: Characklis WG, Marshall KC (eds) Biofilms. John Wiley, New York; 93-130

21   Hamilton WA (1987) Biofilms: microbial interactions and metabolic activities. In: Fletcher M, Gray TRG, Jones JG (eds) Ecology of microbial environments. Cambridge Univ. Press; 361-385

22   Characklis WG, Turakhia MH, Zelver N (1990) Transport and interfacial transfer phenomena. In: Characklis WG, Marshall KC (eds) Biofilms. John Wiley, New York; 265-340

23   Seiferth R, Krüger W (1950) Überraschend hohe Reibungsziffer einer Fernwasserleitung [Surprisingly high drag resistance numbers in a long distance pipeline], VDI-Zeitschr. 92, 189-191

24   Flemming HC (1987) Microbial growth on ion exchangers. Water Res. 21, 745-756

25 Ridgway HF (1988) Microbial adhesion and biofouling of reverse osmosis membranes. In: Parekh BS (ed) Reverse osmosis technology. Marcel Dekker, New York, Basel; 429-481

26 Eckhardt FEW (1985) Solubilization, transport and deposition of mineral cations by microorganisms - efficient rock weathering agents. In: Drever DJ (ed) The chemistry of weathering. D. Reidel Publ.; 161-173

27 Bock E (1987) Biologisch induzierte Korrosion von Naturstein - starker Befall mit Nitrifikanten. Bautenschutz Bautensanierung 10, 24-27

28 Krumbein WE, Lange C (1978) Decay of plaster, paintings and wall material of the interior of buildings via microbial activity. In: Krumbein WE (ed) Environmental biogeochemistry and geomicrobiology. Vol. 2: The terrestrial environment. Ann Arbor Sci., Michigan; 687-697

29 Bielecki R, Schremmer H (1987) Biogene Schwefelsäure-Korrosion in teilgefüllten Abwasserkanälen. Heft 94/1987 Mitt. Leichtweiß-Inst. f. Wasserbau, Univ. Braunschweig

30 Cunningham AB, Bouwer EJ, Characklis WG (1990) Biofilms in porous media. In: Characklis WG, Marshall KC (eds) Biofilms. John Wiley, New York, 1990, 697-732

31 Jacques M, Marrie, TM, Costerton, JW (1987) Review: microbial colonization of prostethic devices. Microb. Ecol. 13, 173-191

32 Colbourne JS, Pratt DJ, Smith MG, Fisher-Hoch SP, Harper D (1984) Legionella and public water supplies. Proc. Int. Conf. Water Wastewater Microbiol., 8.-11. Feb. 1988, Vol 1, 3/1-3/6, Newport Beach, CA

33 Schoenen D, Schulze-Röbbecke R, Schirdewahn N (1988) Microbial contamination of water by pipe and hose materials. 2nd communication: growth of *Legionella pneumpophila*. Zbl. Bakt. Hyg. B 186, 326-332

34 Schulze-Röbbecke R, Schirdewahn N (1989) Mycobacteria in biofilms. Zbl. Bakt. Hyd. B 188, 385-390

35 Müller RC (1946) Medizinische Mikrobiologie. Urban & Schwarzenberg, Berlin, 246

BIOFOULING: EFFECTS AND CONTROL

W.G. Characklis

Center for Interfacial Microbial Process Engineering
Montana State University
Bozeman, Montana
USA 59717

ABSTRACT

Biofouling refers to the undesirable accumulation of a biotic deposit on a surface. The deposit may contain micro- and macroorganisms. The focus of this paper is <u>microbial</u> fouling biofilms which consist of an organic film composed of microorganisms embedded in a polymer matrix of their own making. The composite of microbial cells and EPS is termed a <u>biofilm</u>. The surface accumulation is often composed of significant quantities of inorganic materials. Complex fouling deposits, like those found in industrial environments, often consist of biofilms in intimate association with inorganic particles (1), crystalline precipitates or scale (2), and/or corrosion products (3). These complex <u>deposits</u> often form more rapidly and are more tightly bound than biofilm alone. These deposits are difficult to characterize at the microscale, i.e. at the cellular level.

Thus while biofilm processes, their kinetics, and their stoichiometry can be described in terms of fundamental, intensive variables, this paper must generally describe observations in terms of performance parameters (e.g. heat transfer resistance or fluid frictional resistance).

THE OPERATING PLANT ENVIRONMENT

An industrial operation contains numerous environments where corrosion and fouling processes are potentially troublesome including cooling water systems (recirculating and once-through), storage tanks, water and wastewater treatment facilities, filters, and piping. Microbial fouling and corrosion also occur on ship hulls, reverse osmosis membranes (4), porous media (e.g. groundwater or oil-bearing formation), ion exchangers (5), drinking water distribution systems (6). Biofouling has been reported in turbulent flows and stagnant waters, on smooth surfaces and crevices, and on metals, concrete, and numerous other substrata.

Biofouling frequently occurs in conjunction with other types of fouling including crystalline or precipitation fouling (e.g. scaling) and particulate fouling (frequently due to sedimentation). Microbial activity has been found in calcareous deposits, tubercles, and in deposits of particulate material resulting from sedimentation or adsorption. Increased corrosion rates are frequently reported in these circumstances.

H.-C. Flemming · G. G. Geesey (Eds.)
Biofouling and Biocorrosion in Industrial Water Systems
Proceedings of the International Workshop on
Industrial Biofouling and Biocorrosion, Stuttgart, Sept. 13-14,1990
© Springer-Verlag Berlin Heidelberg 1991

A recirculating cooling tower system provides an illustration of a plant environment in which fouling and corrosion can occur. The evaporative losses of water result in a concentration effect which increases nutrient concentration. The hydraulic residence time, water temperature, and surface area-to-volume are all relatively high. As a consequence, microbial growth rates and cell numbers can be very high. If the fill material is wooden, its degradation rate can be significant due to microbial degradation (fungi) or measures to control microbial activity (chlorine). In heat exchangers, scaling (under certain conditions) and microbial film development on the tubes and areas around flow obstructions can increase heat transfer resistance and (metal) corrosion rates.

FOULING: DEFINITIONS AND DESCRIPTION

Fouling is the formation of deposits on equipment surfaces which significantly decreases equipment performance and/or the useful life of the equipment. Several types of fouling, and their combinations, may occur (7):

1. biological fouling: the accumulation and metabolism of macroorganisms (macrobial fouling) and/or microorganisms (microbial fouling).

2. chemical reaction fouling: deposits formed by chemical reaction in which the substratum (e.g. condenser tube) is not a reactant. Polymerization of petroleum refinery feedstocks is an important example of this type fouling.

3. corrosion fouling: the substratum itself reacts with compounds in the liquid phase to produce a deposit.

4. freezing fouling: solidification of a liquid or some of its higher melting point constituents on a cooled surface.

5. particulate fouling: accumulation on the equipment surface of finely divided solids suspended in the process fluid. Sedimentation fouling is an appropriate term if gravity is the primary mechanism for deposition.

6. precipitation fouling: precipitation of dissolved substances on the equipment surface. This process is termed scaling if the dissolved substances have inverse temperature solubility characteristics (e.g. $CaCO_3$) and the precipitation occurs on a superheated surface.

In most operating plant environments, more than one type of fouling will be occurring simultaneously. For example, microbial fouling is not limited to processes related to biological activity. Microbial fouling also includes the combined result of microbial activity and physicochemical processes in the associated slime layer with the chemical changes at the equipment surface and chemical reactions within the bulk fluid. The interaction can enhance some of the more commonly observed phenomena such as particulate, sedimentation, and corrosion fouling. Because of its complicated composition, the accumulated material will be termed a <u>deposit</u> unless data is available to classify it further.

The <u>interactions</u> between the various types of fouling are poorly understood and, consequently, provide a challenge in diagnosis and treatment.

PROBLEMS CAUSED BY FOULING

Fouling biofilms impair the performance of process equipment. They can form on any surface in contact with a process fluid. The economic consequences of fouling are the essential reason for industrial interest in the fouling of operating equipment. To assess the importance of a fouling situation, the economic and energy penalties arising from the operation of equipment subject to fouling must be considered. The deleterious effects of fouling include the following:

1. energy losses due to increased fluid frictional resistance (e.g. in pipelines, on ship hulls and propellers, and in porous media such as oil and water wells or filters) and increased heat transfer resistance (e.g., power plant condensers and process heat exchangers). Zelver et al. (1) have documented a case of biofouling at a nuclear power plant in which heat transfer rate in a fan cooler decreased by 30% due to biofouling in a 30 day period.

2. increased capital costs for excess equipment capacity (e.g. excess surface area in heat exchangers) to account for fouling. At a Canadian power plant site, biofouling decreased heat transfer rate in a condenser by 30% over a two month period (8). The plant design had allowed for a 15% decrease due to fouling.

3. increased capital costs for premature replacement of equipment experiencing severe under-deposit corrosion. Recently, a nuclear power plant had to replace a condenser after approximately 6 years operation because of severe corrosion attributed partially to microbial activity. The condenser had a design lifetime of approximately 20 years.

4. unscheduled turnarounds or downtime, resulting in loss of production, to clean equipment which fouled at an unanticipated rate. Downtime can cost a power plant as much as $1 million per day because they must purchase the power from elsewhere to serve their customers. In oil production, downtime relates directly to the amount of product being shipped which could extend to $10 million per day.

5. quality control problems resulting from fouling of heat exchange equipment or fouling of product stream (e.g. sliming of paper mills or rolled steel).

6. safety problems. Fouling of service water systems in nuclear power plants is a major concern because it reduces the heat transfer capacity available during an emergency or during an accident. Fouling in drinking water distribution systems may lead to high concentrations of microbes in the drinking water which potentially

affect public health. Deaths due to Legionnaire's disease has been attributed in several instances to fouling in cooling towers.

The anticipated presence of significant fouling can alter the size and other design features of operating equipment. The operation of equipment subject to fouling is constrained by the need to formulate economically justifiable cleaning schedules and internal treatment programs.

An estimate of the economic consequences of fouling was presented by Pritchard (9) for fouling in Britain and suggests the cost was between $600-1,000 million per year (about 0.5% of the British 1976 GNP). Van Nostrand et al. (10) have estimated, that for petroleum refining in the non-Communist countries, the total cost of fouling is $4.4 billion per year. The costs are bound to increase with increasing fuel and material costs. Fouling in heat exchangers may cost the United States billions of dollars annually (11).

In industrial equipment, fouling of surfaces can be the main cause of progressive reduction in performance and efficiency. Accumulation of slime, dirt, and debris in the industrial environment rarely causes concern and, is the cause for the limited attention focused on fouling in the past. However, the cost of problems related to fouling are significant and must be emphasized to designers and operators of equipment.

A RATIONAL APPROACH TO SOLVING FOULING PROBLEMS

Fouling is a complex phenomenon resulting from several processes occurring in parallel and in series. The <u>rate</u> and <u>extent</u> of these processes, in turn, are influenced by numerous physical, chemical, and biological factors in the immediate environment of the substratum. Many laboratory experiments and field observations have resulted in volumes of data without deducing relationships of wide, general use. A conceptual framework for describing fouling processes would be benefical in interpreting available historical data and be invaluable in designing future experimental fouling tests. If the conceptual framework could be stated in mathematical terms, more benefits would accrue including the ability to mathematically simulate fouling processes on the computer, computer "experiments" frequently being less expensive than laboratory experiments. The simulation can be supported by a large database and a logic program to form an "expert" system.

A rational approach, as contrasted with an empirical approach, develops the conceptual framework by resorting to fundamental processes which are reasonably well understood.

A rational approach to fouling entails a process analysis which identifies the contributing fundamental processes and determines the influence of process variables on process <u>rate</u> and <u>extent</u>. The approach requires experimental measurement techniques which permit the elucidation of the specific processes. Many of these fundamental processes have been described mathematically in Characklis et al. (12). The mathematical description of the individual processes can be combined to develop models to extrapolate and generalize experimental results.

The difficulty in generalizing or extrapolating experimental fouling data is related to the complexity of the process which frequently involves heat transfer, mass transfer, momentum transfer, as well as physical, chemical, and biological processes at the surfaces. The goal of a rational approach is to elucidate the fundamental processes that contribute to the overall fouling process. Once these fundamental processes are properly understood, they are incorporated into a mathematical model of the overall fouling process. The model validation requires experiments designed specifically to investigate particular fundamental processes rather than experiments which consider only the overall deposit accumulation process. Such fundamental experiments frequently require more effort than tests which simply observe the overall fouling process, but the fundamental experiments ultimately lead to results that can be applied with greater confidence to a wider range of fouling situations.

The tools of process analysis lead to rational methods for comparing the effects of metallurgy, shear stress, heat flux, water temperature, geometry, biocide treatment, etc., on fouling processes and their influence on equipment performance. Simulation can also be used to test operating and maintenance procedures such as internal treatments and cleaning schedules (13). The process analysis technique may also lead to a more systematic method for developing and evaluating fouling control techniques, regardless of whether they employ physical, chemical, or even biological methods.

Defining Fouling *Rate* and *Extent*

Fouling biofilm accumulation can be considered the net result of the following physical, chemical, and biological processes (14):

1. transport of soluble and particulate components to the wetted substratum,

2. net adsorption of soluble and/or particulates to the wetted substratum,

3. chemical or microbial reaction at the substratum or within the deposit, and

4. net detachment, sloughing, or spalling of portions of the deposit from the wetted substratum.

The overall result is a sequence of events generally characterized by a sigmoidal-shaped progression including three identifiable periods (Fig. 1).

1. an induction period in which very small changes in accumulation are detectable,

2. an exponential increase period which is characterized as the *logarithmic rate* (or *log rate*) of accumulation and is used to represent the kinetics of net accumulation, and

3. an asymptotic or plateau period which is a criterion for the *extent* of net accumulation.

12

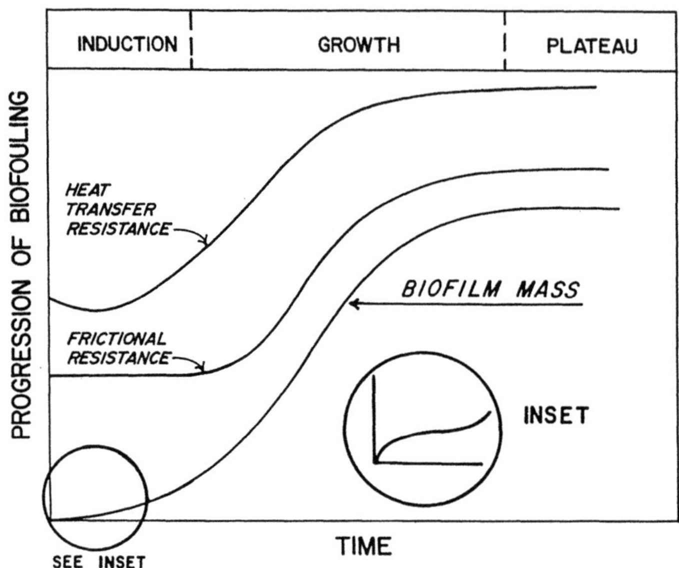

Fig. 1. The progression of biofouling is generally described by a sigmoidal function (Eq. 1) when expressed in terms of deposit accumulation, hydrodynamic frictional resistance, or heat transfer resistance. Three periods or phases can be arbitrarily defined: (i) the *induction* period during which little measurable accumulation occurs, (ii) the *log rate* accumulation period, and (iii) the *plateau* or pseudo steady state period when accumulation is constant. The plateau period represents the *extent* of deposit accumulation

Characklis (14) has described the progression of biofilm accumulation as the result of several distinct processes. Experimental methods and techniques necessary to distinguish the rate of individual contributing processes is sometimes difficult. Yet, characterization of the rate and extent of net accumulation is necessary for evaluation of many field tests in which the more fundamental processes cannot be determined.

The sigmoidal progression can be described by the logistic equation:

$$\frac{dX_f}{dt} = k\, X_f\, (1 - k'\, X_f) \qquad [1]$$

where
X_f = deposit concentration (arbitrary units)
k = rate coefficient (t^{-1})
k' = saturation coefficient (arbitrary units)

Using the logistic equation, the *log rate* of biofouling deposit accumulation can be defined in terms of the rate coefficient k. The *log rate* occurs early in the progression when $k'X_f << 1$. The *extent* of biofouling deposit accumulation can be represented by $X_{f,max}$ or the coefficient k' and is determined at "steady state" when $k' = X_f$. Thus, k and k' (or $X_{f,max}$) are used in this paper as coarse indicators of system kinetics and stoichiometry to describe factors which influence net biofouling deposit accumulation. The terms *log rate* and *extent*, when used in the specific context of net biofouling deposit accumulation, are underlined to emphasize the distinction between these two characteristics of biofouling deposit accumulation. Several variables are used to represent net accumulation and can serve as the variable X in Eq. 1 including biofilm thickness, biofilm mass, fluid frictional resistance, and heat transfer resistance resulting from biofilm accumulation (15).

The rate and extent of fouling can be controlled by many process variables and can vary considerably with location and process environment. Hence, the length of the three periods (induction, exponential increase, plateau) may vary also. For example, the induction period may not be detectable in a water containing very high concentrations of microorganisms since adsorption rate will be very high. On the other hand, a very hard precipitate may form in parallel with the biofilm so that detachment rate is negligible and no plateau period is discernible. The sensitivity of the measurement technique will also influence the length of the various periods. For example, a rather insensitive technique will result in a relatively long induction period.

The deleterious effects of fouling can be experienced at drastically different levels of deposit accumulation. For example, deposition of very few cells in a computer chip manufacturing process can cause significant losses. Approximately 50% of manufactured computer chips are wasted, many because of microbial contamination. In drinking water, on the order of 1,000 to 10,000 coliform cells per m^{-3} are sufficient to disturb the operation of the distribution system. On the other hand, a biofilm thickness of 40 - 60 μm may be necessary to cause an increase in fluid frictional resistance.

Transport and interfacial transfer processes

<u>Transport</u>. Transport of dissolved and particulate materials from the liquid to the wetted substratum is the first step in the overall fouling process and has been described in Characklis et al. (1990a). Transport rates for soluble materials can be predicted reasonably well with accepted empirical relationships. Predicting transport rates of particulates from the bulk fluid to the wetted substratum is more difficult (12).

<u>Deposition</u>. Advective transport has frequently been combined with adsorption-desorption processes into a process termed <u>deposition,</u> primarily because techniques to distinguish between these processes are difficult. Deposition is easier to measure but is the net result of several processes. Frequently, investigators focus on advective transport rates and incorporate a "sticking efficiency" to rationalize measurements of deposition rates (12). Experimental observations, even with inert particulates (16), indicate that sticking efficiency changes as deposition alters the characteristics of the wetted substratum.

14

Advective transport may be the rate-limiting process in some cases of particulate fouling. However, it appears doubtful that advection will limit the rate of deposition for biofouling except in the very early stage of accumulation (i.e. the induction period in Fig. 1) or in very dilute environments.

Detachment. Detachment is a significant process, especially as the deposit becomes thicker and fluid shear stress increases (17). The detached material often deposits in other parts of the system as has been observed with corrosion products (18). Detachment processes are mostly responsible for the asymptotic phase of a typical fouling progression (Figure 1). During this phase, detachment rate is equal to the combined effects of transport, adsorption/adhesion, and reaction on deposit accumulation.

Reactions at the substratum or within the deposit

Chemical and biochemical reactions at the substratum or within the deposit can contribute significantly to the accumulation of the fouling deposit. In some cases, there will be significant interactive effects such as observed with microbially-mediated corrosion which is a combination of microbial growth and/or product formation as well as deterioration of the substratum (e.g. metal dissolution). In this case, microbial (metabolic) products may diffuse to the metal substratum and cause deterioration (corrosion) of the substratum.

Relevant reactions include microbial growth, extracellular polymer formation, endogenous respiration, and death/lysis. Microbial cells, once attached to the substratum, grow, multiply, and form products from chemical energy (chemosynthetic organisms) derived from the bulk water. In most cases, organic compounds (heterotrophs) provide the energy but reduced inorganic compounds (e.g., Fe^{++}, NH_3, $S^=$) can also be used (chemolithotrophs). Photosynthetic organisms can sometimes provide the organic compounds for heterotrophic growth as has been observed in recirculating cooling tower systems. The relative extent to which cells or product (primarily extracellular polymer substances, EPS) dominate the deposit will significantly influence the effectiveness of any chemical treatment in preventing or removing a biological fouling deposit. The biological deposit, an adsorptive surface, will frequently enhance the accumulation of other fouling deposits such as particulate deposits (19), precipitation deposits (20), and corrosion deposits (3).

Measurement and Simulation

Need and purpose
Experimental investigation of the fouling process is a necessary and important route to minimizing its effects on performance. If reliable fouling data can be obtained with full-scale equipment using the actual process fluids, other measurements and simulation are unnecessary. However, many of the parameters of interest vary considerably throughout the equipment and, in addition, vary with time. As a result, interpretation of operating data is difficult and measurements in the laboratory and onsite are necessary so that some measure of control is possible. Laboratory observations under carefully controlled conditions provide the framework for evaluating and interpreting field results where

control of all parameters is not possible. On the other hand, field results indicate inconsistencies in the models and provide the impetus and experimental hypotheses for further laboratory work.

Laboratory tests. Laboratory tests are generally more cost effective than field tests when goals and objectives are concisely defined. The laboratory provides the proper environment for conducting defined tests which lead to useful models in terms of simulating real systems. These tests are a "starting point." Physical, chemical, and biological factors can be controlled at desired levels. Valuable information can be obtained related to design concepts such as influence of alloy (19), fluid velocity, tube geometry (e.g., enhanced heat transfer surfaces), and temperature profiles (21). The effect of water quality (chemical and biological) on these factors can also be evaluated. Operation and maintenance procedures can be evaluated. For example, laboratory tests can evaluate the effectiveness of a treatment (chemical or physical) procedure applied at varying frequency (22, 23). These tests frequently identify the operating conditions which are best for a given treatment procedure. Laboratory experimentation has also resulted in development of novel measurement techniques useful for field tests and monitoring.

Field tests. Ultimately, measurements at the process site are essential. Measurements with fouling monitors simulate the process environment and can be used to evaluate the potential for fouling as well as the effectiveness of treatment programs. Fouling monitors have been used extensively at power plant sites. For example, sidestream fouling monitors are presently capable of qualitatively simulating fouling processes in a condenser. In addition, sidestream monitors are useful for testing the effects of various design, operating, and environmental variables on fouling processes. However, in situ fouling monitors are needed for regulating fouling control programs in the condenser. More sensitive fouling monitors (sidestream and in situ) are needed to detect very thin biofilms (and/ or their effects) and to permit rapid feedback for fouling control treatments. Frequently, on site fouling tests are only conducted for several months starting from a clean tube. There may be slow processes that only manifest themselves in the fouling dynamics after longer periods of time. However, monitors installed *in situ* will supply long term data which can serve adequately for calibrating models.

How do results from fouling monitor tests relate to operation of technical scale equipment? Fouling monitors, at present, are capable of <u>qualitatively</u> simulating fouling processes as they occur in operating equipment. For example, an experimental program using monitors in an onsite sidestream at a nuclear power plant have resolved an important qualitative question: Is the effective chlorine dose in a condenser a product of concentration times duration? The answer is "No". Under realistic (but not necessarily representative) conditions of temperature and flow, over a realistic range of chlorine concentrations and daily exposure, accumulation of a biofouling deposit on titanium tubes is much more dependent on chlorine concentration than on duration of the treatment (13). Accordingly, by implementing higher concentrations and shorter durations, it is possible to achieve a greater degree of fouling control for the same total chlorine dosage (i.e. the product of concentration and duration), within regulatory limits. The essential qualitative result, that there is an optimal combination of concentration and duration (which is almost certainly different from the simple regulatory constraint of 2 hours per day at a dose that just meets the discharge limit), must hold true for operating condensers as well as for the test systems. The test system is similar enough to the operating condenser that there is no reason this should not be true. The experimental program also identified an optimal daily dose schedule under the conditions of the test.

Quantitatively, however, the specific values defining the optimal chlorine dosage (i.e. the most efficient combination of concentration and duration) in an operating condenser is bound to be somewhat different from that identified in a test system at that site, and will doubtless vary with the seasons and the site. Some of the problems related to seasonal variations in fouling are dependent on design of the equipment (e.g. the heat transfer capacity of a condenser) and the change in operation throughout the year (e.g. change in "heat rate", (24)). The least tractable causes of difference from season to season and from site to site are differences in substratum material, water chemistry, nutrient supply, and source microbial species. The variability owing to season and to site conditions can, of course, be taken into account by conducting measurements with a test system at that site and that time. What is not so simply accomplished is the lesser correction between the test system and the operating equipment. The key to the differences between the test system and the operating equipment is scale. For example, an operating power plant condenser tube bundle is fed by a water box, with the result that different tubes experience different temperatures and flow regimes. Each condenser tube is generally an order of magnitude longer than the test section of the fouling monitor. Because of the gradient in heat exchange rate down the condenser tube and because of the nature of the long tube as a plug flow reactor, there will be systematic differences along the length of the condenser tube with respect to wall temperature, water temperature, residual chlorine, nutrient availability, and possible other chemical changes wrought by the upstream deposit. Power plant operations personnel sometimes observe more fouling at the exit of the condenser as compared to the entrance tube region. Thus, a monitored test section can be assembled to mimic any particular short section of the tube, but the same test section cannot mimic the entire span simultaneously. Analysis of fouling deposits from a dismantled nuclear power plant condenser (admiralty brass) indicate, however, that the differences between tubes overwhelmed whatever pattern may have existed in the fouling gradient along the tubes (25).

What limits our ability to transfer conclusions from the fouling monitors to operating equipment? The extrapolation from the monitors to the operating equipment is constrained both by the amount of effort and attention that can be devoted to monitoring in an operational mode (in situ) and by the available scientific information for modifying fouling predictions to accommodate varying environmental or operating conditions. Studies with the fouling monitors at power plants (26) have been conducted in an experimental mode (sidestream) and may require considerable resources for maintaining controlled conditions. An in situ or operational monitoring program places far fewer demands on personnel and other resources as compared to an experimental monitoring program. However, operational monitoring will probably not be as precise as experimental monitoring nor can the operational monitoring be as extensive. Experimental monitoring may identify an optimal chlorination schedule empirically, while an operational monitoring program will only observe responses of fouling rates to treatment because operations will rarely be modified to observe its effect on fouling. Then a modelling effort is required to translate the results from an operational monitoring program into a recommended satisfactory treatment.

Modeling. The objective of developing a model is to determine the coefficients that describe the effects of various operating variables such as temperature, flow rate, and differences in nutrient supply on the dynamics of fouling accumulation. Then, operational monitoring will provide data to calibrate the model for the given application (season and site). This modeling objective is realistic and, with adequate representations of the component processes, will lead to control systems for optimizing biofouling control.

A mathematical model for biofilm accumulation is necessary for relating fouling monitor data to the operating equipment. For example in a power plant condenser, the "plug flow" nature of the condenser tube, the seasonal variation of fouling, and variations of fouling between different tubes in the same condenser are concerns that could be resolved by a validated mathematical model. Field tests have been used to evaluate the influence of physical factors such as substratum composition (e.g. metal alloy) and water velocity on *log rate* and *extent* of fouling. For example, the performance of different alloys has been tested at a power plant site using an instrument which simulates the heat exchange equipment (1). Field tests are also useful in evaluating the performance of contemplated changes in treatment programs and/or process modifications (17, 27). Of all factors influencing fouling, biological variables and chemical water quality appear to be most difficult to simulate and control in the laboratory. Thus, field tests are also necessary.

Measurements related to fouling

Methods for monitoring the progress of the fouling process and methods for characterizing the fouling deposit can be conveniently classified as follows:

1. direct measurement of deposit quantity and/or composition (28).

2. indirect measurement of deposit quantity by monitoring the effects of the deposit on transport processes (e.g. heat transfer and fluid frictional resistance) (15).

Direct measurement includes deposit mass and deposit thickness and are essential for several reasons. Calibration of any indirect method involves comparison with actual quantity of accumulated deposit. Direct measurements are a necessity when using mass conservation equations to establish process relationships. Direct methods are also useful in relating deposit accumulation to fluid frictional resistance and heat transfer resistance.

Indirect methods provide significant benefits including increased sensitivity. For example, organic carbon analysis of biofilm is as much as 25 times more sensitive than biofilm mass measurements (28). In this case, the specific constituent of the deposit (i.e., organic carbon) provides an excellent measure of accumulation. However, if the deposit contains a large amount of silt and sediment, organic carbon will not be representative of

18

the total fouling deposit accumulation. Light absorbance techniques are even more sensitive than carbon analyses but have other limitations (28).

Indirect methods include monitoring the influence of deposits on heat transfer and fluid frictional resistance as discussed above. The extent of the influence of deposits on heat and momentum transfer depends strongly on deposit characteristics (e.g., composition, thermal conductivity, roughness).

FLUID FRICTIONAL RESISTANCE

Fouling deposits cause increased fluid frictional resistance by decreasing the effective diameter of tubes and by increasing the effective roughness of the substratum as described in detail elsewhere (12, 15). Frictional resistance measurement has several advantages as an indicator of fouling:

1. relatively simple and inexpensive

2. in conduit flow, porous media flow, and in ship propulsion, frictional resistance is the quantity of most concern, i.e.,head loss, loss in carrying capacity, clogging, or propulsion efficiency.

However, in some situations, frictional resistance measurements alone are of limited value and sometimes misleading:

1. Frictional resistance measurements in turbulent flow are relatively insensitive until the fouling deposit thickness exceeds a certain value, approximately the thickness of the viscous sublayer. The thickness of the viscous sublayer is inversely proportional to flow velocity in a given geometry.

2. Some deposits such as $CaCO_3$ scale, exhibit a relatively low relative roughness and have a low thermal conductivity. Therefore, frictional resistance will be relatively low even though heat transfer resistance is significant.

3. In some instances, frictional resistance is not the major parameter of interest (e.g. in heat exchangers or condensers, the major concern is heat transfer resistance).

Frictional resistance can increase significantly as a result of even a small increase in roughness. On ship hulls, a equivalent sand roughness height (densely packed) as small as 25 μm can increase drag by 8% while a roughness element of 50 μm will increase drag by as much as 22%. The drag effects become more severe as water velocity increases.

Fouling also causes increased friction losses in porous media by decreasing the effective porosity of a porous media formation, Porous media are dealt with in detail in Cunningham et al. (29).

HEAT TRANSFER RESISTANCE

Combined Heat Transfer and Fluid Friction Measurements

Overall heat transfer resistance is the sum of conductive and advective heat transfer resistance. Advective heat transfer resistance results from fluid motion and generally decreases as the fouling deposit accumulates since the "roughness" of the deposit increases turbulence in the interfacial region. Conductive heat transfer resistance results from insulating layers formed by the deposits and generally increases as the fouling deposit accumulates. The relative changes in advective and conductive heat transfer resistance will depend on the following:

- deposit thickness, deposit roughness, and deposit thermal conductivity
- fluid flow rate
- radial temperature gradient in the tube

Characklis et al. (30) have reported the influence of fouling deposits on conductive and advective heat transfer resistance in tubes in the laboratory and also in plant-scale equipment (12). If deposit thickness is measured along with overall heat transfer resistance and fluid frictional resistance, the effective deposit thermal conductivity and effective deposit roughness can be determined. Effective deposit thermal conductivity and roughness have been determined in several cases by this method (15).

Thus, frictional resistance and heat transfer resistance measurements indicate the effect of fouling on system performance and can indicate the *extent* of fouling but do not yield information on the type of the deposit. At a plant site, the deposits are rarely homogenous but are typically a combination of biofilm, scale, and corrosion products. Results obtained by Characklis et al. (19) indicate significant differences in specific properties (in situ thermal conductivity and relative roughness) between deposits which vary in composition. The different deposit transport properties have been used as a basis for a proposed method for an in situ and non-destructive method for identifying the type of deposit accumulated on the heat transfer surface at a plant site (31).

Heat Transfer Fouling Factor

Heat transfer fouling results are customarily reported in terms of fouling resistance or <u>fouling factor</u> (R_f) which is defined as follows:

$$Rf = U_f^{-1} - U_0^{-1} \tag{2}$$

where R_f = fouling resistance (t^3 T M^{-1})
        U_f = overall heat transfer coefficient for the fouled surface (M t^3 T^{-1})
        U_0 = overall heat transfer coefficient for the clean surface (M t^3 T^{-1})

The use of R_f, fouling resistance, is sometimes misleading and frequently conceals valuable diagnostic information:

1. The influence of fouling on heat exchange rate in engineering design is generally expressed as the thermal (conductive) resistance of the deposit (32). In fact, measured R_f represents the <u>net</u> increase in heat transfer resistance (conductive plus advective resistance) and not thermal (conductive) resistance of the deposit.

2. The conductive resistance of the deposit, in <u>most</u> cases, will be higher than R_f. The extent to which it is greater than R_f will depend on the roughness of the deposit accumulated on the heat transfer surface. The progression of heat transfer resistance due to accumulation of biofilm inside a laboratory tubular reactor experiment (30) was observed to be sigmoidal. In terms of R_f, an increase in heat transfer resistance of 0.00009 m² °C/W was observed, but the increase in conductive resistance was 0.00023 m² °C/W, 2.5 times higher than R_f.

3. The heat exchanger design values for R_f are generally selected from tables of questionable accuracy with vague information as to the operating condition (e.g. shear stress) and the type of deposit (scale, biofilm, etc.) for which R_f values were determined.

4. The calculation of R_f is directly related to the overall heat transfer resistance (i.e., advective resistance) at clean condition (U_0^{-1}). But water velocity influences the advective resistance at clean condition and, hence, calculation of R_f. Thus, R_f at two different velocities are difficult to compare.

5. Most mathematical models (33-35) describing the influence of fouling processes on heat transfer are based on the following relationship:

$$\frac{dR_f}{dt} = r_{dep} - r_{det} \tag{3}$$

where r_{dep} = rate of deposition (t^2 T M^{-1})
           r_{det} = rate of detachment (t^2 T M^{-1})

This relationship expresses the deposition rate and detachment rate of fouling deposit in terms of energy units rather than in term of mass units. The model (Eq. 3) defines R_f as:

$$R_f = L_f / k_{Td} \tag{4}$$

where L_f = thickness of deposit (L)
          k_{Td} = thermal conductivity of deposit (M L t^{-3} T^{-1})

This relationship is <u>only valid</u> when the advective resistance remains constant and R_f equals conductive resistance of deposit. Fouling, in many cases, is accompanied by an increase in pressure drop or decrease in advective resistance.

At constant R_t (i.e. $dR_t/dt = 0$), the deposition rate equals detachment rate (or thickness remains constant as indicated in Eq. 4). However, constant R_t can also result when the increase in conductive resistance (or thickness) of the deposit equals the decrease in advective resistance (due to deposit roughness). The model (Eq. 3) cannot predict a negative R_t. Yet, negative R_t has been observed in numerous studies during initial biofilm accumulation when the increase in conductive resistance (or thickness) of the deposit is _less_ than the decrease in advective resistance. Negative R_t in the early stages of biofilm accumulation has also been observed in the field when new surfaces are initially exposed to a fouling environment. At constant deposit (biofilm) thickness, change in density or composition of the deposit can influence the conductive resistance (R_t) of the deposit. Asymptotic (or steady state) R_t can only result when conductive and advective resistance of the deposit remain constant. However, biofouling deposit properties such as thickness, density, and/or roughness may be varying.

Therefore, R_t may result in ambiguities, especially as related to simplistic models. Nevertheless, R_t has been used to evaluate fouling processes in many field tests when more detailed information was not available. R_t is also used because of the numerous studies which have reported results in those terms.

However, R_t will be used more generally can also include the fluid friction resistance resulting from fouling.

DEPOSIT PROPERTIES

Fouling monitors frequently measure the increase in frictional resistance and/or heat transfer resistance. These measurements indicate the effect of fouling and do not yield information on the type of deposit. Information regarding deposit properties and composition, however, is often useful in selecting an appropriate treatment procedure.

Deposit Mass and Composition

In addition to fouling monitors, analytical methods have been developed to assess the contribution of biological and chemical processes to overall fouling deposit accumulation. As an illustration, results from physical, chemical, and biological analysis of deposits removed from fouling monitors _and_ operating condensers at several power plant sites are illustrated in Characklis (15). Results of deposit analyses from three different sites are presented as examples in Table 1. The differences in deposit composition at different exposure times and between sites are evident.

Table 1 Comparison of deposit characteristics from fouling monitors at three power plant sites. Water velocity in the tubes containing Hudson River water (24) ranged from 1.34 - 2.23 m s^{-1} on a seasonal basis to simulate plant operating schedule. Site B, on an Atlantic estuary, used a water velocity of 2.16 m s^{-1}. Water velocity at the estuarine Canadian plant (8) was 1.97 ms^{-1}.

Exposure Time (days)	Tube Mat'l. (g/m^2)	Deposit Mass (% deposit mass)	Volatile Mass (% deposit mass)	% C	% N	% H
				(% deposit masss)		
			Hudson River			
40	Titanium	1.15	6.28	4.07	0.41	1.27
61	Titanium	6.30	13.45	5.55	0.75	1.35
			Atlantic Estuary			
35	Titanium	14.89	26.65	9.50	0.03	1.43
62	Titanium	37.79	26.70	10.03	1.02	1.85
			Canadian Estuary			
60	Admiralty Brass	7.77	14.07	0.55	2.59	

Deposit Thickness

Deposit thickness is difficult to determine experimentally and has rarely been accomplished in field tests. However, thickness is extremely important because it is required to determine deposit density and distribution of other constituents in the deposit. In addition, the usefulness of deposit roughness depends on the thickness measurement. In some cases, the standard deviation in the thickness measurement represents an estimate of biofilm roughness (36).

Deposit Transport Properties

Industrial fouling deposits are rarely homogenous and can exhibit a wide range of chemical and microbial composition. Not surprisingly then, two deposits of equal thickness can influence heat and momentum transfer in drastically different ways since deposit thermal conductivity and relative roughness can vary widely (Table 2):

Table 2   A comparison of measured values for thermal conductivity and relative roughness of biofilm and $CaCO_3$ deposits (37).

Deposit	Thickness (μm)	Relative Roughness (---)	Thermal Conductivity (W m^{-1} °C^{-1})	Reference
Biofilm	40	0.003		(17)
	165	0.014		(17)
	300	0.062		(17)
	500	0.157		(17)
$CaCO_3$	165[1]	0.0001		
	224[1]	0.0002		
	262[1]	0.0006		
$CaCO_3$			2.26 - 2.93	(38)
$CaSO_4$			2.31	(38)
$Ca_3(PO_4)_2$			2.60	(38)
$Mg_3(PO_4)_2$			2.16	(38)
Fe_2O_3 (magnetic)			2.88	(38)
Analcite			1.27	(38)
Biofilm	$\approx$100		0.63	(19)

[1]Calculated from overall heat transfer resistance assuming a thermal conductivity for $CaCO_3$ of 2.6 W m^{-1} °C^{-1}.

If in situ deposit thickness were determined, effective deposit thermal conductivity and roughness could be determined by measurement of conductive and advective heat transfer resistance, fluid frictional resistance, and engineering correlations. This *in situ* diagnostic method could be incorporated into an on-line fouling monitor. Such a diagnostic tool would find advantage as a feedback control instrument in an operating plant. As more data accumulates, deposit composition could be estimated from thermal conductivity and roughness determinations in much the same way as chemical composition is determined from "libraries" of spectral data associated with gas chromotography-/mass spectrometry.

SUMMARY

Biofouling is the undesirable accumulation of a microbiological deposit on a surface. Complex biofouling deposits found in industrial environments often consist of biofilms in intimate association with inorganic particles, crystalline precipitates, and/or corrosion products. The biofouling process is characterized by a *log rate* and an *extent*, each influenced by a variety of operating and environmental variables including water temperature, fluid velocity, etc.

A rational approach to industrial biofouling processes is necessary for several reasons including the following: 1) to determine the effects of critical variable in any given system, 2) to compare biofouling processes in different systems, 3) to develop standard practices related to design, operation, and control. Such a rational approach is only possible through proficient use of fluid dynamics, energy balances, and mass balances in the operating systems.

ACKNOWLEDGEMENTS

The author gratefully acknowledges the National Science Foundation, Engineering Research Centers Program and the Center for Interfacial Microbial Process Engineering Industrial Associates and Industrial Sponsors for partial support of current research.

REFERENCES

1   Zelver N, Flandreau JR, Spataro WH, Chapple KR, Characklis WG, Roe FL (1982) Analysis and monitoring of heat transfer fouling". ASME, 82-JPGC/Pwr-7

2   Turakhia MH Characklis WG (1988) Activity of P. aeruginosa in biofilms: effect of calcium. Biotech. Bioeng. 33, 406-414

3   Characklis WG, Zelver N, Nelson CH, Lewis RO, Dobb DE, Pagenkopf GK (1984) Influence of biofouling and biofouling control techniques on corrosion of Copper-Nickel tubes. Nat. Assoc. Corrosion Eng. paper no. 250, CORROSION/83, Anaheim, CA

4   Ridgway HF, Justice CA, Whittaker C, Argo DG, Olson BH (1984) Biofilm fouling of RO membranes - its nature and effects on treatment of water for reuse. J. Am. Water Works Assoc. 76, 94-102

5   Flemming HC (1987) Microbial growth on ion exchangers. Wat. Res. 21, 745-756

6   Characklis WG (1988) Bacterial regrowth in distribution systems. Final Report, American Water Works Association Res. Found., Denver, CO

7   Epstein N (1981) Fouling: technical aspects. In: Somerscales EFC, Knudsen EJ (eds) Fouling of Heat Transfer Equipment. Hemisphere Publ. Corp., Washington, DC; 31-53.

8   Characklis WG, Robinson JA (1984) Development of a fouling control program. Final Report, Canadian Electrical Association, No. 219 G 388

9   Pritchard AM (1981) Fouling - Science of art? An investigation of fouling and antifouling measures in the British Isles. In: Somerscales EFC, Knudsen JG (eds) Fouling of Heat Transfer Equipment. Hemisphere Publ. Corp., Washington, DC; 513-523.

10 v. Nostrand WL, Leach SH, Haluska JL (1981) Economic penalties associated with the fouling of refining heat transfer equipment. In: Somerscales EFC, Knudsen JG (eds) Fouling of Heat Transfer Equipment. Hemisphere Publ. Corp., Washington, DC; 619-643.

11 Lund D, Sandu C (1981) Chemical reaction fouling due to foodstuffs. In: Somerscales EFC, Knudsen JG (eds) Fouling of Heat Transfer Equipment. Hemisphere Publ. Corp., Washington, DC; 437-476.

12 Characklis WG, Turakhia MH, Zelver N (1990) Transport and interfacial transfer phenomena. In: Characklis WG, Marshall KC (eds) Biofilms. John Wiley, New York; 265-339.

13 Characklis WG (1990) Microbial biofouling control. In: Characklis WG, Marshall KC (eds) Biofilms. John Wiley, New York; 585-633

14 Characklis WG (1990) Biofilm processes. In: Characklis WG, Marshall KC (eds) Biofilms. John Wiley, New York; 195-231

15 Characklis WG (1990) Microbial fouling. In: Characklis WG, Marshall KC (eds) Biofilms. John Wiley, New York; 523-584

16 Beal SK (1970) Deposition of particles in turbulent flow on channel or pipe walls. Nucl. Sci. Eng. 40, 1-8

17 Characklis WG (1981) Bioengineering report/fouling biofilm development: a process analysis. Biotech Bioeng. 23, 1923-1960

18 Lister DH (1981) Corrosion products in power generating systems. In: Somerscales EFC, Knudsen JG (eds) Fouling of Heat Transfer Equipment. Hemisphere Publ. Corp., Washington, DC; 135-200.

19 Zelver N, Characklis WG, Robinson JA, Roe FL, Dicic Z, Chapple KR, Ribaudo A (1984) Tube material, fluid velocity, surface temperature and fouling: a field study. Cool. Tower Inst., Houston, TX, CTI paper no. TP-84-16

20 Characklis WG, Zelver N, Turakhia MH, Roe FL (1981) Energy Losses in Water Conduits: Monitoring and diagnosis. Proc. 42nd Int. Wat. Conf., Engineers' Soc West Pennsylv., 229-235

21 Characklis WG (1980) Biofilm development and destruction. Final Report, Project RP902-1, Electric Power Research Inst., Palo Alto, CA

22 Norrman G, Characklis WG, Bryers JD (1977) Control of microbial fouling in circular tubes with chlorine. Dev. Ind. Microbiol., 18, 581-590

23 Characklis WG, Trulear MG, Stathopoulos NA, Chang LC (1980) Oxidation and destruction of microbial films. In: Jolley RL *et al.* (eds) Water Chlorination. Ann Arbor Press, Ann Arbor, MI; 349-368.

24 Ferguson RJ (1981) Determination of Seasonal variations in microbiological fouling factors and average Slime thickness. Cool. Tower Inst., Houston, TX, CTI paper no. TP-237A

25 McCaughey MS, Thau A, Characklis WG, Jones WL (1987) An evaluation of condenser tube fouling at an estuarine nuclear power plant. Pres. at Joint Power Gen. Conf., ASME, Miami

26 Garey JF, Jorden RM, Aitken AH, Burton DT, Gray RH (eds)(1980) Condenser biofouling control. Ann Arbor Science, Ann Arbor, MI

27 Matson JV, Characklis WG (1982) Biofouling control in recycled cooling water with bromochlorohydantoin. Cool. Tower Inst., Houston, TX, paper no. TP-250-A

28 Characklis WG (1990) Laboratory biofilm reactors. In: Characklis WG, Marshall KC (eds) Biofilms. John Wiley, New York; 55-89

29 Cunningham AB, Bouwer EJ, Characklis WG (1990) Biofilms in porous media. In: Characklis WG, Marshall KC (eds) Biofilms. John Wiley, New York; 697-732.

30 Characklis WG, Nimmons MJ, Picologlou BF (1981) Influence of fouling biofilms on heat transfer. J. Heat Transf. Eng. 3, 23-37

31 Characklis WG, Zelver N, Turakhia MH, Nimmons MJ (1981) Fouling and heat transfer. In: Chenoweth JM, Impagliazzo M (eds) Fouling of Heat Exchange Equipment. Am Soc Mechan Eng, New York, NY, 1-15

32 Kreith F (1973) Principles of heat transfer. 3rd Ed., Harper and Row, New York; 571-573.

33 Kern DQ, Seaton RE (1959) A theoretical analysis of thermal surface fouling. Brit. Chem. Eng. 4, 258-262

34 Taborek J, Knudsen JG, Aoki T, Ritter RB, Palen WJ (1972) Fouling: the major unresolved problem in heat transfer. Chem.Eng.Prog. 68, 59-67

35 Watkinson AP, Epstein N (1969) Fouling in a gas-oil heat exchanger. Chem. Eng. Prog. Symp. Ser. 65 (92), 84

36 Characklis WG, McFeters GA, Marshall KC (1990) Physiological ecology in biofilms. In: Characklis WG, Marshall KC (eds) Biofilms. John Wiley, New York; 341-394 b

37 Turakhia MH, Characklis WG, Zelver N (1984) Fouling of heat exchange surfaces: measurement and diagnosis. Heat Transf. Engineer. 5, 93-101

38 Sherwood TK, Pigford RI, Wilkie CR (1975) Mass transfer. McGraw-Hill, New York

ROLE OF BACTERIAL ADHESION
IN BIOFILM FORMATION AND BIOCORROSION

K.C. Marshall and Barbara L. Blainey

School of Microbiology
The University of New South Wales
Kensington, NSW 2033 Australia

ABSTRACT

Bacteria are generally small (1 μm or less), negatively-charged bodies with variable cell-surface hydrophobicity and can be regarded as living colloidal particles in relation to their behaviour at surfaces. The process of adhesion of bacteria is considered in terms of their approach to a surface, the effects of long- and short-range forces, and the interactions between bacterial and substratum surface properties. Attached bacteria are capable of metabolizing surface-bound substrates, then they begin to grow in size and reproduce. Different bacteria exhibit various modes of cell division and different mechanisms of detachment of some of the daughter cells. In flowing systems, wherein immobilized bacteria receive a continual supply of nutrients, rapid multiplication and entrapment of additional bacteria result in biofilm development. Such biofilms foul the surfaces of ships, oil rigs, heat exchangers, water reticulation and hydro-electric pipelines, and membrane filter systems. Where metal and concrete surfaces are involved, biofilms play a role in establishing conditions for biocorrosion to occur. Preliminary studies on attempts to prevent or limit bacterial adhesion to surfaces are discussed.

INTRODUCTION

Bacteria in natural habitats normally range in size from 0.15 to 2.0 μm in length or diameter, with the majority at the lower end of this range in oligotrophic habitats (1). Bacteria tend to form stable colloidal suspensions because they have a density only slightly greater than that of water and because they possess a net negative surface charge (2). Various bacteria also differ in the degree (3) and location (4) of cell surface hydrophobicity, although many bacteria are relatively hydrophilic as a result of the production of hydrophilic extracellular polymeric substances (EPS).

Although a consideration of the colloidal properties of bacteria is useful in defining their adhesive capabilities, it must be stressed that bacteria are not inert colloidal particles. They are living organisms and, as such, are capable of metabolism, growth and, in many instances, independent mobility. The outer surface of the cell, in intimate contact with the external environment, varies considerably between different bacteria and even in the same bacterial type under different growth conditions (5). A detailed discussion of the chemistry of the cell envelope and extracellular components of Gram-positive and Gram-negative bacteria in relation to bacterial adhesion processes, is presented by Wicken

H.-C. Flemming · G. G. Geesey (Eds.)
Biofouling and Biocorrosion in Industrial Water Systems
Proceedings of the International Workshop on
Industrial Biofouling and Biocorrosion, Stuttgart, Sept. 13-14,1990
© Springer-Verlag Berlin Heidelberg 1991

(6). The colloidal properties of bacterial cells do give some insights into the behaviour of bacteria at surfaces, but it is also necessary to consider the biological responses of the bacteria to the microenvironment at the solid-water interface.

TRANSPORT OF BACTERIA TO SURFACES

Fluid Dynamic Forces

Currents in water bodies provide the major mechanism for the transport of bacteria over large distances. Eddy diffusion in turbulent flow systems tends to disperse bacteria resulting in a uniform concentration in the bulk water phase (7), but near a surface the bacteria are transported to the region of the viscous (or boundary) layer. When the bacteria are travelling at a greater velocity than the water in the viscous layer, then a lift force directs the bacteria towards the surface where frictional drag forces slow them down (7). In fast flowing systems, turbulent downsweeps (8) also direct bacteria towards the surface.

Sedimentation

Bacterial sedimentation is of significance mainly in quiescent waters and only with aggregates (flocs) of many bacteria. In general, bacteria form stable suspensions and only sediment when the suspension is destabilised following aggregation by a variety of mechanisms (9).

Chemotaxis

A chemotactic response to any nutrient gradient occurring near a surface obviously can only be demonstrated in motile bacteria. In turbulent water conditions it is unlikely that a nutrient gradient would be established, but in the viscous layer near a surface such gradients could exist and motile bacteria would then approach the surface by swimming in the direction of the nutrient gradient.

Brownian Motion

Small bacteria (less than 1.0 μm diameter) exhibit a significant degree of Brownian displacement, and such motion could account for random contacts by small bacteria with surfaces in quiescent water (2).

SUBSTRATUM SURFACE PROPERTIES AND CONDITIONING FILMS

The adhesion of bacteria to apparently inert surfaces is influenced both by the inherent properties of the substratum surface and the way in which these properties are modified by molecular adsorption at the surface (conditioning films).

Substratum Properties

Any solid surface may be characterized in terms of its surface charge, surface free energy, and surface roughness. *Surface charge* depends on the balance between the numbers of positive and negative charges exposed at the solid surface, *surface free energy* depends, in part, on the relative degree of hydrophobicity or hydrophilicity of the surface components, and *surface roughness* will vary considerably in different substratum materials, but even apparently smooth surfaces may possess undulations with an amplitude in excess of 1.0 μm - sufficient to provide protection to small bacteria in the turbulent flow systems.

Conditioning Films

The surface charge and surface free energy of a substratum surface are modified on immersion into a natural habitat as a result of the spontaneous adsorption of macromolecules and smaller hydrophobic molecules at the solid-liquid interface. Alterations in *surface charge* may be detected by measuring the electrophoretic mobility of particles in a microelectrophoresis apparatus. For instance, Neihof and Loeb (10) demonstrated that particles with different surface charge properties exhibited a very narrow range of negative electrophoretic mobilities after exposure to seawater. However, exposure of the particles to seawater treated by UV irradiation to destroy organics resulted in normal electrophoretic mobilities of the particles. The adsorption of organic molecules to the particle surfaces immersed in the natural seawater clearly altered the surface charge of the substratum surface.

Alterations in the *surface free energy* following immersion of a substratum surface may be determined by measuring the contact angles made by a series of liquids of defined surface tension following drying of the exposed surface (11) or by measuring contact angles of air bubbles at the substratum-liquid interface (12). Significant alterations in the surface free energy of certain substrata are readily demonstrated following immersion of the surfaces in natural habitats or in solutions of pure proteins or other molecules.

Modification of the surface properties of substrata by conditioning films does alter the degree of bacterial adhesion to such surfaces (12, 13). In addition, the adsorbed organic molecules often serve as substrates for bacterial metabolism and growth (see appropriate sections later in this chapter).

REVERSIBLE ADHESION

Early Definition

Marshall et al. (14) defined *reversible adhesion* as an instantaneous attraction of bacteria to a surface in such a manner that the bacteria continued to exhibit Brownian motion and could be removed from the surface by a moderate shear force (e.g., a jet of water or by the bacterium's own motility). The concept of reversible adhesion has proved useful in studies on bacterial adhesion and should continue to be so in the future. Some aspects of the definition and of the attempts to explain the phenomenon of reversible adhesion certainly require further consideration.

Reversible Adhesion Reconsidered

In view of developments since the original definition of reversible adhesion was proposed, the following points need to be emphasized:

Instantaneous attraction

Bacteria in the vicinity of a solid-liquid interface certainly exhibit an instantaneous attraction to the interface (14), but this is not exclusively reversible adhesion as some bacteria are able to adhere irreversibly immediately on contact with the interface (15). As a corollary of this, the concept of irreversible adhesion being a time-dependent function must be revised and will be considered in more detail in the next section.

Brownian motion

A normal rod-shaped bacterium (e.g. a pseudomonad) exhibits Brownian motion when reversibly adhering and lacks such motion when irreversibly adhering (14). With bacteria that adhere in a perpendicular orientation, such as *Flexibacter* and *Hyphomicrobium* (4), the extended portion of the cell can still exhibit Brownian motion following irreversible adhesion. As a result, it would seem that any mention of Brownian motion in a generalized definition only confuses the issue.

DLVO theory

An attempt was made by Marshall et al. (14) to explain the reversible nature of bacterial adhesion by means of the colloid stability (or DLVO) theory. This theory accounts, at least in part, for the attraction of a negatively- charged bacterium to a negatively-charged substratum surface at the secondary attraction minimum, resulting from the interaction between London-van der Waals attraction energies and electrical repulsion energies in the overlapping double layers surrounding the negatively charged surfaces. Direct observations revealed that attraction of *Achromobacter* R8 to glass surfaces increased as the monovalent and divalent electrolyte concentration increased as might be predicted by the DLVO theory (14). Similar results have been reported by other authors (16-18).

Problems in the application of the DLVO theory to biological systems have been raised, especially when the complexity of the bacterial cell envelope and the extracellular materials are taken into consideration (19, 20). Recently, however, Busscher and Weerkamp (21) have argued strongly in favour of the role of the secondary attraction minimum in the initial attraction of bacteria to surfaces.

Polymer bridging

Assuming that a bacterial cell is held at a finite distance from a surface (say 10 nm) by repulsion forces, then bridging polymers (such as pili, fimbriae, other proteins, polysaccharides) probably provide the major mechanism for the firm adhesion of bacteria to surfaces (14). It is feasible that the difference between reversible and irreversible adhesion is a matter of the degree of polymer bridging. Few polymer bridges may be readily broken by a mild shear force, whereas extensive bridging makes the bacterium more resistant to a shear force. Again, this possibility requires further investigation.

Thermodynamic aspects

Busscher et al. (22) have considered the reversibility of adhesion of streptococci to solids in terms of interfacial free energies of adhesion (DFadh). Using several streptococci and a range of different substrata of widely varying surface free energies, these authors found that adhesion was reversible where DFadh is positive. Thus, the hydrophobic Streptococcus mitis T9 adhered reversibly to hydrophilic glass surfaces, whereas the hydrophilic S. sanguis adhered reversibly to more hydrophobic substrata but adhered to glass irreversibly.

Metabolism and growth of reversibly adhering bacteria on surface localized substrates

The marine bacterium, *Vibrio MH3*, is unable to adhere irreversibly to surfaces, but is attracted reversibly to surfaces and is capable of scavenging surface-bound stearic acid (23). A more recent study (24) revealed that small starved cells of *Vibrio MH3* grew to normal size adjacent to a surface with bound stearic acid as the sole substrate, began the process of cell division, and moved away from the surface just prior to the separation of daughter cells. These results certainly show that bacteria capable only of reversible adhesion are able to benefit from nutrients concentrated at the solid-liquid interface.

IRREVERSIBLE ADHESION

Early Definition

Irreversible adhesion was originally defined as a time-dependent firm adhesion whereby the bacteria no longer exhibit Brownian motion and could not be removed by a moderate shear force (14). As indicated in the previous section, the question of the presence or absence of Brownian motion following irreversible adhesion depends on the length and mode of orientation of the bacterial cell at the substratum surface. Other features of the process of irreversible adhesion require elaboration.

Irreversible Adhesion Reconsidered

The nature of irreversible adhesion probably varies according to both the organism and the substratum involved in the adhesion process.

Time dependency

The time-dependent component of irreversible adhesion of bacteria to surfaces was envisioned by Marshall et al. (14) as the time required by the bacterium to respond physiologically, by producing bridging polymer, to the presence of a surface. Evidence supporting this suggestion has been obtained (Blainey and Marshall, unpublished results) using tritiated-thymidine labelled *Pseudomonas EK20* exposed for two hours to polystyrene petri dish surfaces prior to determining the percentage of cells either (a) unattached (in the supernatant), (b) loosely attached (washed from the surface after pouring off the supernatant), or (c) firmly attached to the surface. The supernatant and washings were then reexposed to fresh petri dish surfaces for a further two hours and the various categories of attachment again estimated. An additional control was the original bacterial suspension exposed to petri dishes for four hours. The cells in the supernatant, which had little or no prior exposure to a surface, exhibited similar adhesive behaviour to those in the original test suspension when both sets were incubated for two hours (Table 1). The cells in the washings, which had been closely associated with a surface in the original incubation, showed a very high level of firm adhesion suggesting the "switching on" of some adhesive responses during that time. Longer exposure to the surface of the original suspension gave a higher level of firm attachment; an observation that is consistent with a "switching on" of an adhesive response with time.

Table 1. The effect of prior exposure of *Pseudomonas EK20* to a polystyrene surface on its ability to firmly adhere

Test suspension	Assay time (h)	% of population		
		Firm attachment	Loose attachment	Unattached
Original	2	15	57	27
2 h supernatant	2	15	58	26
2 h washings	2	76	12	12
Original	4	40	43	16

Other studies have revealed that some bacteria exhibit an essentially instantaneous irreversible adhesion to surfaces (15). A good indication of the effect of substratum surface properties on the mode of adhesion is the finding (12) that *Pseudomonas NCMB2021* displays an essentially instantaneous irreversible adhesion to a hydrophilic tissue culture dish polystyrene substratum but shows a time-dependent, and ultimately a much greater degree of, irreversible adhesion to a hydrophobic petri dish polystyrene substratum. Differential removal of this organism from these substrata by proteolytic enzymes (12) suggested that different adhesion mechanisms operated on each substrata. This suggestion was confirmed for the adhesion of *Vibrio proteolytica* to hydrophilic and hydrophobic surfaces (25). Short-range forces: Extracellular polymers of bacteria possess a small radius of curvature and, consequently, they can overcome the repulsion barrier near a surface and be deposited at or near the primary attraction minimum (19). At this point, they are subjected to a variety of short-range forces, namely:

(i)   chemical bonds, such as electrostatic, covalent, and hydrogen bonds,
(ii)   dipole interactions, such as dipole-dipole (Keesom), dipole-induced dipole (Debye), and ion-dipole interactions, and
(iii)   hydrophobic interactions (19, 20).

The distinctive mechanisms of adhesion exhibited by a bacterium at various substrata may result from the production of different polymers or they may originate in changes in the configuration of a single polymer at the different substratum surfaces resulting in various combinations of short-range forces becoming operative.

Thermodynamic approach

Following some confusing results regarding the relative importance of substratum surface free energy in bacterial adhesion, it was considered necessary to account for the effects of bacterial surface free energy and liquid surface tension in addition to the substratum surface free energy in the adhesion process (26). The change in free energy associated with bacterial adhesion (ΔFadh) is given by:

$$\Delta\text{Fadh} = \gamma\text{BS} - \gamma\text{BL} - \gamma\text{SL}$$

where γBS, γBL, and γSL are the interfacial tensions of the bacterium-substratum, bacterium-liquid, and substratum-liquid, respectively. According to the thermodynamic model, bacterial adhesion is favoured if the process results in a free energy decrease. In general, good agreement was found between bacterial adhesion to a variety of substrata and the adhesion behaviour predicted by the thermodynamic model (25).

This work needs to be extended to include bacteria with a wider range of relative hydrophobicities and hydrophilicities, as well as attempts to determine the modifying effects of conditioning films on substratum surface free energy and, as a result, on the adhesion process.

DETACHMENT OF BACTERIA FROM SURFACES

Careful observations of bacteria adhering to surfaces have revealed a variable number of bacteria detaching from the surfaces. Several explanations for the detachment of previously irreversibly adhering bacteria have been proposed.

Changes in Bacterial Surface Properties

The production of capsular polysaccharides was found to interfere with the adhesion of *Acinetobacter calcoaceticus* to hydrocarbons by masking the adhesive fimbriae (27). In a similar manner, hydrophobic benthic cyanobacteria have been shown to produce a hydrophilic capsule that results in release of the cyanobacteria from surfaces (28).

Changes in Substratum Surface Properties

Power and Marshall (24) observed a slow form of motility and, ultimately, detachment of cells of *Pseudomonas JD8* on a surface treated with stearic acid. This phenomenon was explained by a change from a hydrophobic to a hydrophilic substratum following utilization of the stearic acid leading to reversible adhesion as described by Busscher et al. (22), followed by movement to a site where hydrophobic stearic acid was present leading again to irreversible adhesion, and then repetition of this process.

Metabolic State of Bacteria

Bright and Fletcher (29) reported detachment of bacteria from surfaces as a result of the bacteria being in a less active metabolic state than those remaining on the surface. This condition may result in altered surface properties of the bacteria.

Release of Daughter Cells

Bacteria adhering in a perpendicular manner at surfaces release daughter cells directly to the aqueous phase (30), as described for Vibrio DW1 where a single attached cell was shown to produce successively four daughter cells before the mother cell was lost from the surface.

Polymer Cleavage

Polymers binding cells to surfaces may be disrupted chemically (31), enzymatically (12), by removal of calcium ions (32), or by shear forces in fast flowing systems.

Whatever the mechanism of detachment, the process results in the release of bacteria to the aqueous phase, sometimes in significantly large numbers. The process is particularly important where potentially pathogenic organisms are released from the surfaces of domestic water pipeline systems (33).

STARVATION SURVIVAL

Oligotrophic Habitats

Most natural microbial habitats are oligotrophic. An oligotrophic habitat has been defined by Poindexter (34) as one with a nutrient flux of from near zero to a fraction of 1 $mgl^{-1}d^{-1}$ of C_{org}. Copiotrophic bacteria, that require relatively high nutrient fluxes for cell growth and reproduction, are found in quite large numbers in most oligotrophic habitats. What are copiotrophic bacteria doing in such habitats and how do they maintain these numbers over long time periods? Even when organic substrates appear in oligotrophic habitats, as a result of autolysis or other forms of breakdown of living organisms, or as a result of upwelling processes in ocean environments, the copiotrophic bacteria will rapidly utilize these substrates and reestablish the oligotrophic environment. Once the readily available substrate is consumed, copiotrophic bacteria tend to starve.

Starvation-Survival

The ability of copiotrophic bacteria to adapt to starvation conditions in order to ensure survival until sufficient energy substrate for growth appears has been termed starvation-survival (35). In the original studies on this process by Novitsky and Morita (36-

38), it was observed that copiotrophic bacteria responded to starvation conditions by dramatic reductions in cell size and endogenous respiration, and it was suggested that such responses provided a strategy for the continued survival of the bacteria under non-growth conditions.

Detailed studies of changes in cell volume following transfer of exponential phase *Vibrio DW1* into a starvation medium (39, 40) have revealed a short period, within five hours, where cell numbers increase and cell volumes decrease rapidly. This period results from a pre-programmed cell division cycle in the bacteria and has been termed the fragmentation period to distinguish it from cell division leading to active cell reproduction (40). Continued starvation leads to a further reduction in cell volume without any change in total cell numbers. Amy and Morita (41) have reported a number of different starvation-survival patterns following studies on a range of marine bacteria. Not all bacteria fragment on transfer to a starvation medium and the loss in viability varies considerably in different bacteria subjected to starvation conditions. Recent studies on starvation-survival of marine bacteria have been reviewed by Kjelleberg et al. (42).

METABOLISM OF AND GROWTH ON SURFACE-BOUND SUBSTRATE

Initial Surface Colonization

Early investigations on the initial colonization of surfaces immersed in marine habitats indicated that the primary colonizing organisms were very small bacteria and these were followed by a succession of larger rods and then by prosthecate bacteria (43). Later it was realized that the small bacteria were probably starved bacteria and, in the presence of nutrients concentrated at the surface, these small bacteria may grow to normal size giving the appearance of a secondary population colonizing the surface (44). If this was so, then growth and reproduction of starved bacteria at surfaces would constitute a source of organisms being released into the aqueous phase and would explain the presence of significant numbers of copiotrophic bacteria in oligotrophic habitats. Consequently, our attention has been focussed on demonstrating such a cycle under controlled laboratory conditions.

Bacterial Scavenging of Surface Bound Stearic Acid

Kefford et al. (45) proposed the use of bound stearic acid as a model to demonstrate scavenging of a substrate from a surface. A monolayer of stearic acid was formed on the surface of water and the monolayer transferred to the surface of a glass cylinder by dipping the cylinder through the stearic acid layer on the water. A hydrophobic, wild type *Serratia marcescens* rapidly adhered to the surface and quickly utilized some stearic acid prior to the rate of utilization falling away rapidly. By comparison, a hydrophilic, non-pigmented mutant of *S. marcescens* did not adhere well and only slowly utilized a small amount of stearic acid. A reversibly adhering *Leptospira* utilized the fatty acid at a moderately fast rate which continued until most of the stearic acid bound to the surface was exhausted. Because the motile leptospire was able to return to the aqueous phase, cells of this organism harvested from the aqueous phase showed significant accumulation of ^{14}C-label from the stearic acid. Little such accumulation was found when *S. marcescens* was harvested in the same way.

Because the leptospire is a very unusual organism in terms of its morphology, motility, and growth, Hermansson and Marshall (23) decided to deliberately isolate a marine organism that failed to adhere irreversibly to surfaces. This was successfully achieved and it was found that the organism, *Vibrio MH3*, was capable of reversible adhesion and of scavenging surface-bound stearic acid. Thus, bacteria are capable of utilizing surface-bound substrates irrespective of whether they are reversibly or irreversibly bound to the surface. Presumably, the bacteria near the surface utilize the low levels of fatty acid in solution and alter the equilibrium so that a continual supply of fatty acid slowly becomes available.

Cellular Growth and Reproduction on Surface-Bound Substrate

In order to directly visualize cellular growth and reproduction, we have followed these phenomena using bacteria adhering to a dialysis membrane in a Duxbury (46) microperfusion chamber. Continuous observations were made using oil-immersion, phase contrast optics linked to a television camera, time-synch generator, and time-lapse video recording equipment. The medium perfused through the chamber could be completely lacking energy substrate, contain oligotrophic levels of energy substrate, or high levels of such substrate. Bacteria were starved prior to addition to the dialysis membrane.

Kjelleberg et al. (30) reported a study of the growth and reproduction of starved cells of *Vibrio DW1* in an essentially oligotrophic medium. The organism did not grow in the aqueous phase of this medium, yet vigorous growth was observed at the membrane surface. *Vibrio DW1* adhered to the surface in a perpendicular manner and the small, starved cells were observed to grow to normal size, begin the cell division process, release a daughter cell into the aqueous phase, the attached mother cell would elongate again, and repeatedly release daughter cells into the aqueous phase. In a rich medium, this organism exhibited a generation time of about 37 minutes. At a surface with the oligotrophic medium, the average time between successive cell divisions was 57 minutes. This result certainly suggests that nutrients concentrated from the perfusate onto the membrane surface in sufficient quantities to maintain this substantial growth rate. When the membrane surface became relatively saturated with adhering bacteria, the cells stopped growing and reverted to small starvation forms because the number of cells exceeded the rate of accumulation of substrate from the medium.

In a study with *Pseudomonas JD8*, the perfusate contained no substrate and the only source of energy for the organism was stearic acid bound to the dialysis membrane surface. Again starved, adhering cells were observed to increase to normal size and to undergo binary fission at the membrane surface. The cells of *Pseudomonas JD8* adhered in a face-to-face orientation and exhibited very unusual behaviour following cell division (24). When the daughter cells separated, they continued to migrate away from each other across the surface at an extremely slow rate (0.15 0.08 m min-1). This slow migration was explained by Power and Marshall (24) in terms of the cells being irreversibly attached to the hydrophobic stearic acid covered surface but, upon utilization of the stearic acid in the microzone around the cell, the cells became reversibly attached to the now hydrophilic substratum (22) and could move. As soon as the cells moved a short distance, they encountered more stearic acid (hydrophobic) and irreversibly adhered until that substrate was utilized and then the cycle repeated. The supply of fatty acid at the surface was limited and, when most of it was exhausted, many cells became detached from the surface as predicted by Busscher et al. (22).

Power and Marshall (24) also examined the behaviour of the reversibly adhering Vibrio MH3 on a stearic acid coated membrane surface. Small, starved cells of this organism also grew to normal size and began the process of cell division near the surface. At this point, the cells detached from the surface and, apparently, completed the cycle of cell division in the aqueous phase (out of focus) because daughter cells were found to return to the surface where they grew and, again, began the division cycle.

It is obvious that starved bacteria, whether reversibly or irreversibly attached to a surface, have the potential to utilize surface bound substrates for metabolism, growth, and reproduction. In addition, different organisms possess different growth strategies and different mechanisms for ensuring the release of cells into the aqueous phase.

BIOFILM FORMATION

In flowing water systems, where continual replenishment of energy substrate is assured, extensive growth of primary and secondary colonizing bacteria occurs leading to biofilm formation. A biofilm consists of cells immobilized at a substratum and frequently embedded in an organic polymer matrix of microbial origin (47). It is not necessarily uniform in time or space and may be composed of a significant fraction of inorganic or abiotic substances held together by the biotic matrix (47).

The formation of a biofilm can be considered in three stages (48):

1 Biofilm formation begins with a progressive coverage of the surface by multiplication of the primary colonizing bacteria and further colonization by bacteria from the aqueous phase. This stage is characterized by low population densities, a low level of competition between the colonizing organisms, and organisms with relatively high growth rates (r-strategists, 49).

2 A transition stage with multilayers of cells becoming embedded in their own polymer material. This results in increasing population density, increasing competition, and a progressive development of populations with higher competitive abilities (K- strategists, 49).

3 The development of a mature biofilm, in which the population density is high, with high levels of competition, and where K-strategists dominate.

Community Interactions in Biofilms

A range of gradients form in biofilms and result in the development of diverse populations with increasing depth of the biofilm (47). Organisms near the biofilm- water interface possess the advantage of maximum access to oxygen, nutrients and, in exposed biofilms, light to ensure their active growth (Fig. 1).

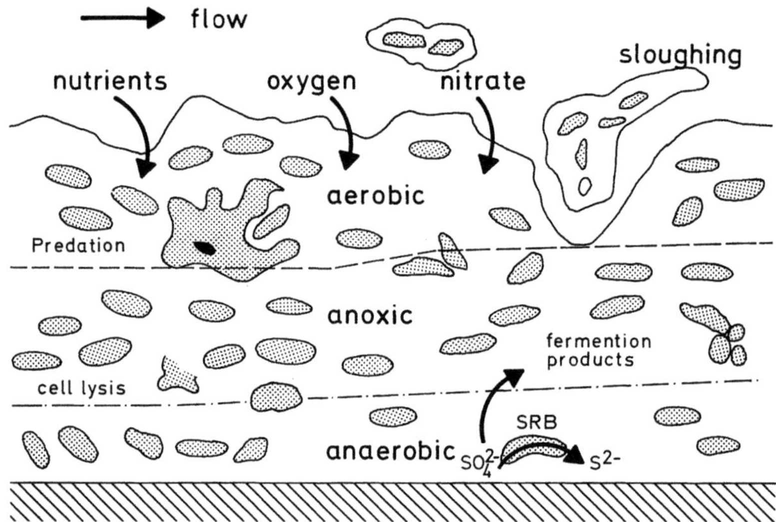

Fig. 1 A schematic representation of a mature biofilm showing various gradients and community interactions

Organisms at intermediate levels within biofilms are subjected to greater competition for oxygen and nutrients and range from microaerophiles, to bacteria utilizing alternative terminal electron acceptors (e.g. nitrate), and to oxygen tolerant anaerobes. At the base of mature biofilms conditions are anaerobic and favour the activities of fermentative and sulphate-reducing bacteria (SRBs). Autolysis of moribund cells adds to the fermentation products as nutrients available within the biofilm. The activity of SRBs in biofilms on metallic surfaces can lead to microbial corrosion (50).

Amoebae function as effective predators within biofilms, whereas ciliates, flagellates, and rotifers actively graze at the biofilm-water interface. It is likely that grazing of the biofilm surface by these micropredators helps to maintain the biofilm community in an actively growing state. Growth within biofilms, however, generally provides bacteria with a remarkable degree of protection from antibacterial agents such as chlorine, surfactants, antibiotics, bacteriocins, bacteriophage, and predatory organisms (51).

PREVENTION OF BACTERIAL ADHESION

Antifouling paints based on the slow release of toxic heavy metals (copper, tin) have become a target of environmentalists because of detrimental effects on aquatic organisms. In addition, many microorganisms exhibit a remarkable degree of tolerance to high levels of heavy metals and tend to colonize treated surfaces and form biofilms, thus negating the effects of the antifouling paints.

An alternative approach is to prevent the adhesion of bacteria to a surface. Owens et al. (52) have described the use of a water-soluble copolymer, Synperonic F108 (I.C.I., UK) of poly(ethylene oxide) and poly(propylene glycol) that inhibited bacterial adhesion to hydrophobic surfaces under conditions of defined hydrodynamic shear. We have examined the effects of a range of these copolymers on the adhesion of pure cultures to polystyrene surfaces (Blainey and Marshall, unpublished data) and have confirmed the overall inhibition of adhesion by the copolymers (Fig. 2). The effectiveness of the inhibition appears to be related to the length of the hydrophilic moiety of the copolymers (Fig. 2). It is felt that the hydrophilic chains bind water in an ice-like structure and prevent polymer bridging to the surface by bacteria (52).

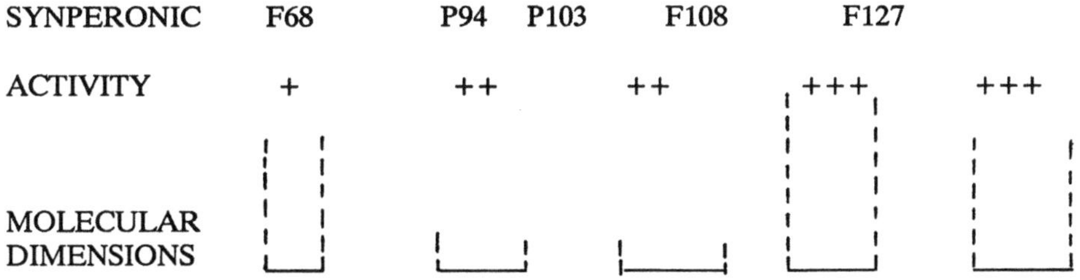

SYNPERONIC	F68	P94	P103	F108	F127
ACTIVITY	+	+ +	+ +	+ + +	+ + +

MOLECULAR DIMENSIONS

Fig. 2 Schematic representation of various Synperonic copolymers and their relative activity in inhibiting adhesion of Pseudomonas EK20 to polystyrene. _____ hydrophobic, - - - - hydrophilic moieties of copolymers; P, paste, F, flakes. Scale: $\overline{MW}$ of 1,000 = __

Table 2. Effectiveness of a dimethylamide in inhibiting bacterial adhesion to or removal of bacterial biofilms from polystyrene surfaces

Organism	Hydrophobicity (%)	Inhibition of adhesion	Removal of biofilm
SW8	98	+	+ +
S9	88	+	+
SW7	78	-	-
SW1	77	+	+/-
EK20	21	-	+/-

Testing the Synperonic F108 in marine habitats revealed a partial inhibition in the short term (1-2 days) but no effect over a period of several weeks (Blainey and Marshall, unpublished results). This long-term failure of the copolymer may result from removal from the surface by displacement or abrasion, masking by adsorption of macromolecules from the natural waters, biodegradation, or the ability of some bacterial exopolymers to penetrate the protective barrier created by the copolymer hydrophilic chains.

Recently it has been shown that *N,N*-dimethylamides of 18 carbon unsaturated carboxylic acids (e.g. Busperse 47, Buckman Laboratories) are useful in controlling biofilms in cooling water systems. These compounds tend to form a monomolecular film on hydrophobic surfaces because of the hydrophobic characteristics of the fatty acid portion of the molecule (53). We have tested Busperse 47 for its ability to inhibit bacterial adhesion to polystyrene and to remove preformed biofilms from polystyrene surfaces. The results (Table 2) show variable effects with different bacteria, as no correlation was observed between the relative hydrophobicity of the bacteria employed and the degree of inhibition of adhesion or removal of biofilm (Blainey and Marshall, unpublished data).

Further studies on strategies to prevent bacterial adhesion are under consideration.

REFERENCES

1   van Es FB, Meyer-Reil L-A (1982) Biomass and metabolic activity of heterotrophic marine bacteria. In: KC Marshall (ed) Advances in Microbial Ecology, Vol. 6, Plenum Press New York; 111-170

2   Marshall KC (1976) Interfaces in Microbial Ecology. Harvard Univ. Press Cambridge MA

3   Rosenberg M, Kjelleberg S (1986) Hydrophobic interactions: role in bacterial adhesion. In: KC Marshall (ed) Advances in Microbial Ecology, Vol. 9, Plenum Publ. Co. New York; 353-393

4   Marshall KC, Cruickshank RH (1973) Cell surface hydrophobicity and the orientation of certain bacteria at interfaces. Arch. Microbiol. 91, 29-40

5   Knox KW, Hardy LN, Markevics LJ, Evans JD, Wicken AJ (1985) Comparative studies on the effect of growth conditions on adhesion, hydrophobicity, and extracellular protein profile of *Streptococcus sanguis G9B*. Infect. Immun. 50, 545-554

6   Wicken AJ (1985) Bacterial cell walls and surfaces. In: DC Savage and M Fletcher (eds) Bacterial Adhesion: Mechanisms and Physiological Significance. Plenum Press New York; 45-70

7   Characklis WG (1981) Bioengineering report: Fouling biofilm development. A process analysis. Biotech. Bioeng. 23, 1923-1960

8　Lister DH (1981) Corrosion products in power generating systems. In: EFC Summerscales and JG Knudsen (eds) Fouling of Heat Transfer Equipment. Hemisphere Washington, DC; 135-200

9　Harris RH, Mitchell R (1973) The role of polymers in microbial aggregation. Ann. Rev. Microbiol. 27, 27-50

10　Neihof R, Loeb G (1974) Dissolved organic matter in seawater and the electric charge of immersed surfaces. J. Mar. Res. 32, 5-12

11　Baier RE (1980) Substrate influence on adhesion of microorganisms and their resultant new surface properties. In: G Bitton and KC Marshall (eds) Adsorption of Microorganisms to Surfaces. Wiley-Interscience New York; 59-104

12　Fletcher M, Marshall KC (1982) Bubble contact angle method for evaluating substratum interfacial characteristics and its relevance to bacterial attachment. Appl. Environ. Microbiol. 44, 184-192

13　Fletcher M (1976) The effects of proteins on bacterial attachment to polystyrene. J. Gen. Microbiol. 94, 400-404

14　Marshall KC, Stout R, Mitchell R (1971) Mechanisms of the initial events in the sorption of marine bacteria to surfaces. J. Gen. Microbiol. 68, 337-348

15　Fletcher M (1980) The question of passive versus active attachment mechanisms in non-specific bacterial adhesion. In: RCW Berkeley, JM Lynch, J Melling, PR Rutter and B Vincent (eds) Microbial Adhesion to Surfaces. Horwood Chichester; 197-210

16　Orstavik D (1977) Sorption of Streptococcus faecium to glass. Acta Pathol. Microbiol. Scand. B 85, 38-46

17　Jones GW, Richardson LA, Uhlman D (1981) The invasion of HeLa cells by *Salmonella typhimurium*: Reversible and irreversible bacterial attachment and the role of bacterial motility. J. Gen. Microbiol. 127, 351-360

18　van Loosdrecht MCM, Lyklema J, Norde W, Zehnder AJB (1989) Bacterial adhesion: a physicochemical approach. Microb. Ecol. 17, 1-15

19　Tadros TF (1980) Particle-surface adhesion. In: RCW Berkeley, JM Lynch, J Melling, PR Rutter and B Vincent (eds) Microbial Adhesion to Surfaces. Horwood Chichester; 93-116

20　Rutter PR, Vincent B (1980) The adhesion of microorganisms to surfaces. In: RCW Berkeley, JM Lynch, J Melling, PR Rutter and B Vincent (eds) Microbial Adhesion to Surfaces, Horwood Chichester; 79-92

44

21 Busscher HJ, Weerkamp AH (1987) Specific and non-specific interactions in bacterial adhesion to solid substrata. FEMS Microbiol. Rev. 46, 165-173

22 Busscher HJ, Uyen HM, Weerkamp AH, Postma WJ, Arends J (1986) Reversibility of adhesion of oral streptococci to solids. FEMS Microbiol. Lett. 35, 303-306

23 Hermansson M, Marshall KC (1985) Utilization of surface localized substrate by non-adhesive marine bacteria. Microb. Ecol. 11, 91-105

24 Power K, Marshall KC (1988) Cellular growth and reproduction of marine bacteria on surface-bound substrate. Biofouling 1, 163-174.

25 Paul JH, Jeffrey WH (1985) Evidence for separate adhesion mechanisms for hydrophilic and hydrophobic surfaces in *Vibrio proteolytica*. Appl. Environ. Microb. 50, 431-437

26 Absolom DR, Lamberti FV, Policova Z, Zingg W, van Oss CJ, Neumann AW (1983) Surface thermodynamics of bacterial adhesion. Appl. Environ. Microbiol. 45, 90-97

27 Rosenberg E, Kaplan N, Pines O, Rosenberg M, Gutnick D (1983) Capsular polysaccharides interfere with adherence of *Acinetobacter calcoaceticus* to hydrocarbons. FEMS Microbiol. Lett. 17, 157-160

28 Fattom A, Shilo M (1984) Hydrophobicity as an adhesion mechanism of benthic cyanobacteria. Appl. Environ. Microbiol. 47, 135-143

29 Bright JJ, Fletcher M (1983) Amino acid assimilation and electron transport system activity in attached and free-living marine bacteria. Appl. Environ. Microbiol. 45, 818-825

30 Kjelleberg S, Humphrey BA, Marshall KC (1982) Effects of interfaces on small, starved marine bacteria. Appl. Environ. Microbiol. 43, 1166-1172

31 Marshall KC (1973) Mechanisms of adhesion of marine bacteria to surfaces. In: RF Acker, BF Brown, JR DePalma and WP Iverson (eds) Proc. 3rd Intern. Congr. Mar. Corrosion Fouling, Northwestern Univ Press Evanston; 625-632

32 Turakhia MH, Cooksey KE, Characklis WG (1983) Influence of a calcium- specific chelant on biofilm removal. Appl. Environ. Microbiol. 46, 1236-1238

33 LeChavallier MW, Cawthon CD, Lee RG (1988) Factors promoting survival of bacteria in chlorinated water supplies. Appl. Environ. Microbiol. 54, 649-654

34 Poindexter JW (1981) Oligotrophy: fast and famine existence. In: M. Alexander (ed) Advances in Microbial Ecology, Vol. 5, Plenum Press New York; 63-89

35 Morita RY (1982) Starvation-survival of heterotrophs in the marine environment. In: KC Marshall (ed) Advances in Microbial Ecology, Vol. 6, Plenum Press New York; 171-198

36 Novitsky J, Morita RY (1976) Morphological characteristics of small cells resulting from nutrient starvation of a psychrophilic marine vibrio. Appl. Environ. Microbiol. 32, 617-622

37 Novitsky J, Morita RY (1977) Survival of a psychrophilic marine vibrio under long-term nutrient starvation. Appl. Environ. Microbiol. 33, 635-641

38 Novitsky J, Morita RY (1978) Possible strategy for the survival of marine bacteria under starvation conditions. Mar. Biol. 48, 289-295

39 Dawson MP, Humphrey BA, Marshall KC (1981) Adhesion, a tactic in the survival strategy of a marine vibrio during starvation. Curr. Microbiol. 6, 195-198

40 Kjelleberg S, Humphrey BA, Marshall KC (1983) Initial phases of starvation and activity of bacteria at surfaces. Appl. Environ. Microbiol. 45, 978-984

41 Amy PS, Morita RY (1983) Starvation-survival patterns of sixteen freshly isolated open-ocean bacteria. Appl. Environ. Microbiol. 45, 1109-1115

42 Kjelleberg S, Hermansson M, Marden P, Jones GW (1987) The transient phase between growth and nongrowth of heterotrophic bacteria, with emphasis on the marine environment. Ann. Rev. Microbiol. 41, 25-49

43 Marshall KC, Stout R, Mitchell R (1971) Selective sorption of bacteria from seawater. Canad. J. Microbiol. 17, 1413-1416

44 Marshall KC (1979) Growth at interfaces. In: M Shilo (ed) Strategies of Microbial Life in Extreme Environments, Verlag Chemie Weinheim; 281-290

45 Kefford B, Kjelleberg S, Marshall KC (1982) Bacterial scavenging: Utilization of fatty acids localized at a solid-liquid interface. Arch. Microbiol. 133, 257-260

46 Duxbury T (1977) A microperfusion chamber for studying the growth of bacterial cells. J. Appl. Bacteriol. 43, 247-251

47 Characklis WG, Marshall KC (1990) Biofilms: A basis for an interdisciplinary approach In: WG Characklis and KC Marshall (eds) Biofilms. Wiley-Interscience New York; 3-15

48 Marshall KC (1989) Growth of bacteria on surface-bound substrates: Significance in biofilm development. In: T Hattori, Y Ishida, Y Maruyama, RY Morita, A Uchida (eds) Recent Advances in Microbial Ecology, Japan. Scientific Soc. Press Tokyo; 146-150

49 Andrews JH, Harris RF (1986) r- and K-selection and microbial ecology. In: KC Marshall (ed) Advances in Microbial Ecology. Vol. 9, Plenum Press New York; 99-147

50 Hamilton WA (1985) Sulphate-reducing bacteria and anaerobic corrosion. Ann. Rev. Microbiol. 39, 195-217

51 Costerton JW, Cheng K-J, Geesey GG, Ladd TI, Nickel JC, Dasgupta M (1987) Bacterial biofilms in nature and disease. Ann. Rev. Microbiol. 41, 435- 464

52 Owens NF, Gingell D, Rutter PR (1987) Inhibition of cell adhesion by a synthetic polymer adsorbed to glass shown under defined hydrodynamic stress. J. Cell. Sci. 87, 667-675

53 Lutey RW, King VM, Cleghorn MZ (1989) Mechanisms of action of dimethylamides as a penetrant/dispersant in cooling water systems. 50th Ann. Intern. Water Confr. Pittsburg

BIOFOULING IN WATER TREATMENT

H.C. Flemming

Institut für Siedlungswasserbau, Wassergüte- und Abfallwirtschaft
der Universität Stuttgart
Bandtäle 1, D-7000 Stuttgart 80

"The organism always wins"
(K.C. Marshall, very unpublished comment on the prospects of biofouling control)

ABSTRACT

Biofouling is understood as the unwanted deposition and growth of living organisms on surfaces. In water treatment, in almost all cases it is caused by microorganisms. They can contaminate the water, cover and block surfaces, host pathogens, and attack their support. Biofouling is a biofilm problem. It is defined operationally and refers to that extent of biofilm growth which interferes with the demands of water production or consumption process. Control of biofouling requires effective detection. Water samples give no valuable information to localize biofilms or to assess their extent. Thus, surfaces have to be investigated. Simple field methods are presented as well as laboratory techniques to evaluate the presence of biofilms. Sanitization of biofouling has to include the removal of biofilms rather than the killing of all cells. Sanitization strategies usually have to break the physical stability of the biofilm matrix by chemicals **and** to remove the biofouling layer by shear forces. Prevention of biofouling depends on a "clean system philosophy": clean equipment, raw water and chemicals, early detection of biofilm formation and early cleaning measures. In most systems, biofilm development can only be prevented with high expenditures. Thus, the extent of biofilm accumulation has to be kept below the level of interference. The extent of biofilm growth is the result of the balance between growth and detachment and it is dependent on the nutrient situation, the temperature and the shear forces. The control of the nutrient situation in the fluid phase can be much more effective than the control of the cell numbers in order to reduce biofilm thickness. For the extent of biofilm accumulation as well as for cleaning measures, the stability of the biofilm matrix is the crucial factor. It depends on the shear forces, the nutrient situation, stabilizing filaments and particles and destabilizing chemicals and internal processes. "Coexistence with biofilms" requires a continuous awareness and strategies similar to those with which some marine animals prevent microbial colonization on their surface.

BIOFOULING PHENOMENA

Biofouling is referred to as the unwanted deposition of biological matter from the bulk liquid to any surface of water treatment, storing or transport equipment. In this chapter, it shall be restricted to the adsorption, accumulation and growth of microorganisms on these surfaces.

H.-C. Flemming · G. G. Geesey (Eds.)
Biofouling and Biocorrosion in Industrial Water Systems
Proceedings of the International Workshop on
Industrial Biofouling and Biocorrosion, Stuttgart, Sept. 13-14, 1990
© Springer-Verlag Berlin Heidelberg 1991

48

In production processes which depend on the water quality, biofilms can cause severe interferences, when microorganisms are released from these biofilms and return to the bulk liquid phase (1). The production of microelectronics, for example, is highly sensitive to particles in the water which is used to rinse the printed circuits (2; see chapters 2 and 7). The sources of the bacteria in the process water are:

a) cells transported with the raw water, air, and chemicals (these cells can be filtered and are susceptible to disinfection)
b) cells introduced by the equipment of the plant and by personnel
c) cells released from biofilms (which are poorly susceptible to microbicides; see chapter 6).

While abiotic particles may be controlled by sophisticated filter systems, microorganisms are particles which can multiply. Even very few microorganisms, which eventually penetrate a filter, can attach to surfaces of the water distribution system, multiply there and seed the water phase with more microorganisms that can attach to printed circuits, cause shortcuts and expensive production losses.

Thus, most of the microorganisms in high purity water systems come from biofilms rather than from breakthroughs through particle filters or filtration membranes. In pharmaceutical systems, strong efforts are necessary to avoid microbial contamination of products, even if the raw water is extensively pretreated (3).

It may be surprising to encounter microbial growth in high purity water, such as 18 MOhm water used in microelectronic industry (4; see chapter 7) with practically no detectable organic Carbon. However, microorganisms can grow in presence of extreme low nutrient concentrations (5; see Table 1 in chapter 1). Some starving cells are known to change their morphology and their surface characteristics, exhibiting a higher hydrophobicity and adhesiveness (6). Once some cells are attached to a surface in a high purity water system, they may scavenge nutrients and concentrate them locally. Considering the small size and mass of bacteria (10^{-12} g per cell (7), one million cells make up a mass of 1 μg, 90 % of which consist of water. Thus, 10^6 cells contribute approximately 0,1 μg of C_{org} which is close to or below the detection limit of most methods. Given enough time (e.g. a few years of the operation time of a plant), this biofilm will grow and provide nutrients for more demanding cells which pass the filters and can attach to the biofilm, feeding on the dead oligotrophic "pioneers". Thus, it is possible to find microorganisms in high purity water systems, which need relatively high concentrations of organic carbon. This can be explained by the local accumulation of scavenged nutrients from the bulk liquid. Thus, even water with very low concentration of nutrients may provide a nutrient-saturated environment for growth and reproduction in the biofilm if its volumetric flow rate is high (8).

Almost every surface in contact with non sterile water can carry a biofilm. In membrane technologies such as reverse osmosis and ultrafiltration (chapter 5), membrane surfaces can be colonized, resulting in a fouling layer. This leads to an increased energy consumption, concentration polarization, losses in mineral rejection, flux decline and in many cases to the contamination of the product water (9). When porous media (10) such as ion exchangers (11, 12), filters (13) or activated carbon adsorption reactors (14-16) are subject to biofouling, an increased fluid resistance up to total clogging, capacity loss and microbial contamination of the treated water are the consequences. Point-of-use water treatment devices can be colonized (17), again, resulting in contamination of the treated water.

The surfaces of degazifiers frequently carry biofilms, scavenging additional nutrients from the air. Geller (18) showed that the nutrient uptake from the air represents a significant factor supporting microbial growth in nutrient limited systems.

The inner surfaces of water pipelines can carry biofilms which increase the drag resistance, leading to more than 50 % loss of transport performance (19, 20) release microorganisms to the bulk liquid, and cause episodes of elevated bacterial numbers in drinking water (21). Even high shear stress systems can be colonized by biofilms (8). Experiments with *Pseudomonas atlantica* in a shear stress gradient showed that a value of about 120 dynes/cm^2 represented a treshold for the attachment (22).

A particular problem is represented by the fact that biofouling usually occurs together with other kinds of fouling (such as the deposition of minerals, organics and colloids; see chapter 2).

In medicine, biofilms can colonize implanted devices of all kinds (23, 24), resulting in health hazards, which are enhanced by the fact that biofilm bacteria are much more resistant to the bodies infection defense (25).

Drinking water systems

Drinking water systems are frequently colonized by biofilms which can contaminate the water (26). LeChevallier et al. (27) examined a drinking water facility in New Jersey where bacterial counts were above the permitted levels. Sections of pipes were removed for biofilm analysis and pipe scrapings were taken to assess the relationship between biofilms within the water distribution system and the quality of the water produced. The treatment plant was not the main source of bacteria as coliform counts from the plant were below federal guidelines. Higher coliform counts were detected in the pipe distribution systems. Also, other heterotrophic bacteria were detected at much higher levels in both plant effluent and at different sample sites. Analysis of the bacterial species showed that the majority of the isolates were present in both the water and in the biofilms. The authors considered that some bacteria must have detached from the biofilm and increased the counts in the pipeline sections. Treatment with 1.0 mg/l free chlorine did not control the problem as the bacteria in the biofilm were protected from the disinfectant.

The distribution of biofilms can be quite patchy (21) with large areas of a surface free of microbes. In an extensive study, Ridgway and Olson (28) report that, though distributed sparsely and randomly along the pipe surface, a variety of morphologically distinguishable bacteria-like structures were observed. The iron-oxidizing bacterium *Gallionella*, recognized by its characteristic helical stalks, was observed both in water samples and attached to pipe surfaces. Large numbers of rod-shaped bacteria were also evident on the surfaces of detritus or silt particles recovered from water samples by filtration. The

50

encrustations of drinking water pipes carry frequently high numbers of bacteria (29), especially coliforms (27, 30) and high chloride levels (21). The film often contains organic or inorganic debris from external sources. Inorganic particles may result from the adsorption of silt and sediment, precipitation of inorganic salts, or corrosion products (21). O'Connor and Banerji (31) found that the microbial growth on pipe wall was related to total organic carbon (TOC) levels in the water and that TOC levels of 5,0 mg/l or less limited the amount of microbial growth in their laboratory system. Very little growth was observed at an organic-carbon level of 0,5 mg/l. A typical TOC level for treated groundwater is 0,5 to 3,0 mg/l while the TOC level in treated surface water is in the range of 2,0 to 5,0 mg/l (21).

Van der Wende et al. (32, 33) and LeChevallier et al. (27) found that biofilm growth and detachment of biofilm cells accounted for most, if not all, the *planctonic* cells present in the bulk water of a chlorine free system. Edyvean (13) emphasizes that such biofilms may have significant effects on filtration, fluid flow, heat transfer, corrosion and the development of and release of pathogens and corrosive substances into the system. Mycobacteria are frequently present in biofilms (34, 35), even in hot water systems of hospitals (36). There are reports on microbial growth on expansion joints of reservoirs (37) and other surfaces used in drinking water treatment (38), such as polyester insulations of reservoirs (39). The growth of *Legionella pneumophila* in biofilms on stainless steel, copper, glass, polyvinylchloride, polyamide, polytetrafluorethylene, silicon and rubber materials of pipes and hoses in water systems is reported (40). After 72 h, a biofilm completely covered the surface of PVC plastic tubes through which drinking water was flowing (41), with microorganisms embedded in thick deposits of extracellular material (42).

Air/water interfaces

Biofilms can also exist at the gas-water interface, where both hydrophobic microorganisms and molecules accumulate (43, 44) and the bacteria can utilize hydrophobic substances as nutrients (45, 46; see chapter 1). When the water level of a storage tank is lowered, this biofilm can attach to the tank wall. It will continue to grow submerged when the water level rises again, and contribute to the microbial contamination. Apart from this mechanism, it can be dislodged into the bulk fluid and distributed by eddies (47), and contribute to episodic high cell counts in drinking water.

Effects of biofouling

In summary, biofilms can affect the

* Water quality (contamination by released microorganisms)
* Hydrodynamic parameters (biofilms are viscoelastic (48) and consume kinetic energy; biofilms can clog porous media (10)
* Substratum properties (biofilms block membranes; turn hydrophobic surfaces into hydrophilic; make surfaces slippery; change the colour; cause biocorrosion or biodeterioration - see chapters 8-11)

The consequences of biofouling are very costly:

* Decrease in product quality
* Loss in plant performance
* Increase in cleaning expenses and down time
* Damage and shorter life of plant components
* Security and health problems for both personnel and environment

The costs are enormous (see chapter 2). An example: the US Navy assessed the costs for the additional fuel demand to move its ships, which is caused by the increased drag resistance by biofilms, to $ 500.000.000 per year (White, pers. comm.). Pumping costs in water pipelines because of the above mentioned increase of the drag resistance, fouling of heat exchanger surfaces and water treatment equipment can easily be summed up to extremely high amounts of money every year.

The effects of biofouling on plant performance, water quality and the equipment are frequently enhanced because they are not recognized in time, underestimated and linked with wrong causes. Thus, they require expensive and difficult countermeasures, which could have been avoided if the problem would have been recognized earlier. That is why awareness is a "preventive tool" in biofouling and biocorrosion control.

The general problems associated with the treatment of biofouling can be categorized as follows:

1. Detection of biofilms
2. Sanitization of a biofouled system
3. Prevention of biofouling
4. Keeping biofilm accumulation *below* an action limit

DETECTION OF BIOFOULING

Process parameters

In very many cases, operational parameters (such as quantity of produced water, pressure drop, filter resistance etc.) are not sufficiently sensitive to detect biofouling at an early stage. In this stage, the parameters will not indicate a biofouling problem, although there may be a biofilm already present. The effects of biofouling will be detected even less readily when the equipment is operating only at a fraction of its maximum capacity (8).

Thus, first there is a biofilm accumulating and second there is "biofouling", when the biofilm leads to undesired consequences, identified as of microbiological origin. "Biofouling" is an operational definition: *biofouling is that extent of biofilm accumulation and activity which interferes with process demands*. In order to understand biofouling problems, it is necessary to understand the properties and dynamics of biofilms.

52

Experiments on the colonization of reverse osmosis membranes have shown (49) that the colonization of the surface occurs very rapidly with the highest adhesion rates during the first hour (fig. 2), until a primary plateau is reached after a few hours of contact. In this period, many of the process parameters are not yet stable, thus, the effects of primary colonization are easily masked. It is interesting that the same adhesion velocity is observed with bacteria which have been killed by peracetic acid compared to living cells (50). This indicates that the adhesion process in this system is a passive one, contrary to an active adhesion, in which the cells produce adhesive substances in response to contact with the surface (51). In practice, this means that killing of suspended microorganisms is not enough - some of them may stick to the surface even if they are dead, because the "glue" is a part of their cell wall. These cells, when attached to surfaces, may provide both substrate (nutrients) and substratum (solid surface for growth) for subsequent cells arriving with the raw water (see "sanitization").

Biofilm accumulation usually follows a sigmoid curve (shown schematically in Fig. 1), as already described in chapter 2:

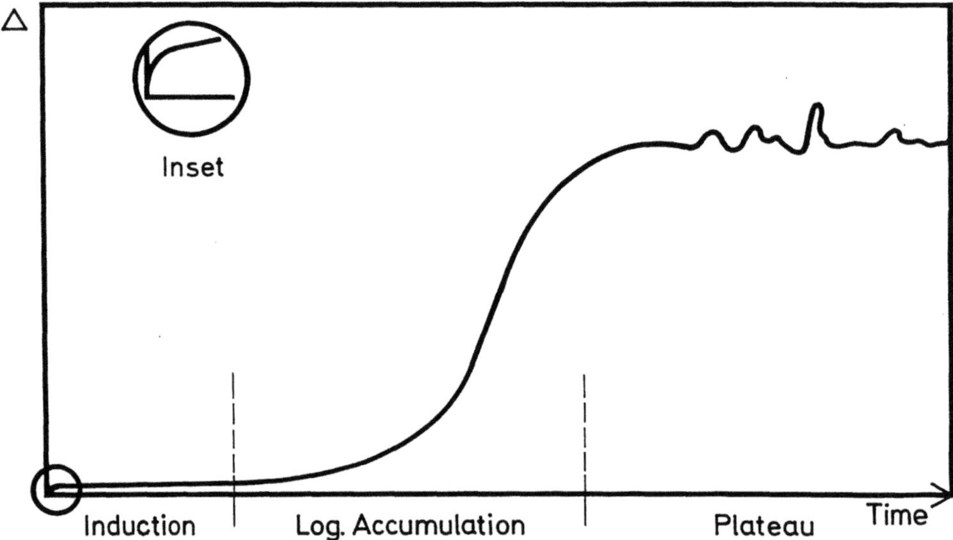

Fig. 1. Progression of biofilm accumulation (after Characklis [1], modified). Δ = biofilm accumulation parameter (thickness, mass, cell number etc.)

The inset in Fig. 1 indicates that the accumulation process begins very early and there is a thin biofilm already during the induction phase (Fig. 2).

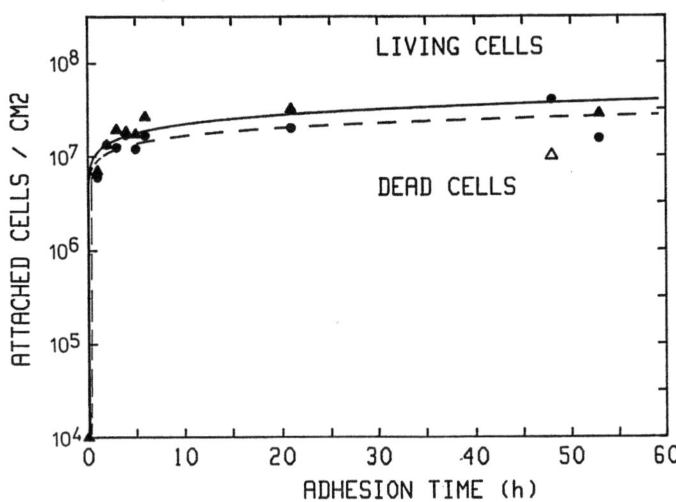

Fig. 2. Primary colonization of polyamide membranes by *P. diminuta*. ●—● = living bacteria, ▲----▲ = dead bacteria (Flemming & Schaule [49], modified)

Cell numbers in the bulk

As already mentioned, cell numbers in the bulk can not be correlated with the extent, location and viability of biofilms. Biofilm cells are released not in a proportional way and suspended cell numbers allow no assessment of the sessile fraction of the microorganisms. Low numbers do not mean that there is no biofilm - in systems with high water flow, biofilm cells can be diluted below the detection limit. Biofilms on materials which leach nutrients (52) seem to seed the bulk liquid more than biofilms on "inert" materials such as stainless steel, although the cell numbers in the biofilms do not differ significantly (21). Nevertheless, many measures to detect and control biofouling are still based on suspended cell numbers, probably, because they can easily be determined.

The surface approach

The bacterial density in a typical slime on a RO surface is between 10^6 and 10^8 per cm² (53, 54). This exceeds the number of cells per ml in tested waters in most cases by several orders of magnitude. Therefore, only a direct approach to surfaces gives information about biofilms, if one wants to know about an actual biofouling situation. Thus, the importance of monitoring sessile bacteria numbers is emphasized for good reasons (55). Allen et al. (29) stated that bacterial enumeration by conventional methods (scraping of the biofilm from the surface followed by bacterial counts) greatly underestimates the true bacterial populations present, e.g. in tubercles. Certainly, viable counting techniques based on colony growth on solid media only recover a fraction of the living cells, especially when mixed populations are enumerated (21). Of course, this has to be considered in the interpretation of the data.

In most practical situations, one will encounter biofilms which have reached the plateau phase. At this stage, some very simple field methods and observations can give indications for biofilms, such as:

- visual inspection of surfaces, especially of crevices, fittings, valves etc.

- slimy layer on accessible surfaces, detectable by wiping with a finger or a clean tissue

- "pyrolytic field test" to get an indication about the biological nature of a fouling layer: remove a small part of the layer, roast it on a lighter's flame and check if it smells of burnt hair

- A macroscopic method using specific dyes is under development (Flemming & Schaule, in prep.).

Biofilms consist mainly of microorganisms, extracellular polymer substances (50-90% of the total biofilm mass), entrapped particles and dissolved substances (48, 54, 56).

The further identification of biofouling is based on laboratory investigations. These include usually one or all of the following methods:

* microscopic inspection (epifluorescence)
* cultivation of microorganisms
* analysis of the deposit
    - water
    - organic matter (TOC)
    - protein
    - carbohydrates
    - pyrogens (Novitsky, 1987)
    - fatty acids
    - ATP
    - activity of various enzymes

Biofilms are characterized by:

* high water content (70-95 %)
* high content of organic matter (50-95 % of dry matter)
* high numbers of colony forming units (CFU) and cells (microscopic counts)
* high content of carbohydrates and proteins
* high content of ATP
* (in most cases) low contents of inorganic matter

In some cases, the result may not be totally consistent with this general description, because frequently other kinds of fouling are involved. A biofilm may change the properties of a given surface (e.g. a hydrophobic polymer) and accelerate the adsorption of particles such as humic substances, clay and debris from the raw water. This material may influence both physical and physiological properties of the biofilms as well as the element composition.

SANITIZATION OF A BIOFOULED SYSTEM

General considerations

The first measurement to sanitize a fouled system is usually the application of a biocide. However, it is a common fallacy that in a system suffering from biofouling, the problem is solved by killing all the microorganisms present. To eliminate the bacteria circulating in the system does not mean you control the biofilm.

It is well known that biofilms are much more resistant against biocides than suspended bacteria (57-65; chapter 6), protected by means of their EPS. The microorganisms in the central layers of the biofilm are most protected (66). Dead ends, corners and fissures may bear biofilms which are not adequately penetrated by biocides. Surviving cells in biofilms at these places act as an inoculum for the cleaned system, explaining the rapid after-growth which is frequently reported (9, 67). The spectrum of organisms that survived tre-atment processes includes spore-formers, acid-fast bacilli, pigmented organisms, disinfec-tant-resistant bacterial strains, various yeasts, fungi and actinomycetes (25, 57, 58, 60, 61). Copper surfaces are well colonized by microorganisms (Jolley et al., 1988). Biofilms have been found even on the walls of the copper pipes of a disinfectant distribution system (58). Thus, the suspended bacteria may be dead, but some biofilm bacteria still survive. Again, sampling of the bulk water phase is misleading, if one wants to evaluate the effec-tiveness of a sanitizing measure, because the situation of the biofilm can not be recogniz-ed this way.

Additionally, in many cases even a dead biofilm may cause problems, as is the case in reverse osmosis (68) or in porous media. It can accelerate the regrowth of biofilms, because it provides both substratum and substrate for bacteria subsequently transported with the raw water. Costerton (pers. comm.) assesses that dead biofilms impede the re-growth about ten fold. Thus, for sanitization it may be much more important to remove the biofilm completely rather than to kill it and leave it on its place.

The Sanitization Concept

Sanitization of the feedwater is part of the sanitization programme. It should result in a decrease of the cell numbers and, even more important, a decrease of nutrients. Characklis (8) has shown that the adhesion of new cells is almost negligible compared to biofilm growth due to nutrients. The nutrient level of the feedwater offers a tool to control biofilm thickness, although it will not lead to the complete disappearance of bio-films.

The strategy has to address the biofilms and may include chemical measures which decrease the mechanical stability of the biofilm matrix and physical measures (Tab. 1)

The effectiveness of biocides depends on a number of intrinsic and environmental factors, such as the kind of biocide, the biocide concentration, the biocide demand (which can be decreased by prior cleaning), interference with other dissolved substances, pH, Temperature, the contact time, types of organisms present, their physiological state and, most important, the presence of biofilms.

As a general rule, the higher the temperature, the longer the contact time and the higher the concentration of the disinfectant, the greater will be the degree of disinfection. Considering the fact that a biofilm is increasingly difficult to remove the older it is (54), early action may prove economical.

It should be emphasized that biocides are toxic materials and every biocide added to the system can cause environmental problems. Restrictive environmental legislation may prohibit the application of an agent in effective concentrations without special water treatment.

Physical methods

In most cases, the sanitization programme will include the weakening of (i) the biofilm matrix and (ii) of the adhesion strength to the supporting surface by chemicals prior to the application of shear stress by flushing.

Physical methods to clean a biofouled system can be summarized as in Table 1:

Table 1. Some Physical methods to clean biofouled surfaces

Method	comments
Flushing	simplest method; limited efficacy biofilms thinner than the viscous sublayer are not sheared (69) Special case: flushing supported by cleaners and/or after application of chemical agents which destabilize the biofilm matrix and adhesion to the support
Backwashing	effective for loosely adherent films in tubes, on filters, to a certain extent in ion exchangers, but not below a certain level (70)
Air bumping	very limited efficacy (70, 71)
Abrasive sponge balls	demonstrated efficacy, but possible problems because of the abrasion of protective oxide films (72)

Table 1, continued

Non-abrasive sponge balls	extensively used in industry; problems with thick biofilms and with smearing organics
Sand scouring	difficult to control abrasive effects
Brushing	very effective, but limited applicability, expensive; can lead to the selection of firmly adhering species (73)
Hot water, steam	used in high purity water systems in Japan with good results (2); saves expensive and possibly harmful and toxic chemicals. Hot water systems may select for thermophiles and are reported to carry biofilms including mycobacteria (36)
Ice nucleating	application of ethylene glycol at -12°, destabilizes the biofilm matrix and detaches it from the support (74); no practical experiences reported
Irradiation	very low effectivity against biofilms; entrapped particles and opaque biofilms may shield bacteria (75); biofilms on protection mantles of UV irradiators have been reported (76; Chrtek & Flemming, unpubl. obs.)
Ultrasonic energy	promising method for soft biofilms; application limited to non-sensitive material; some biofilms are extremely stable (77, 78)

Whittaker et al. (53) and Yanagi and Mori (79) report in comparative studies that mechanical cleaning gave the best results in terms of biofilm removal from reverse osmosis membranes. This method is obviously limited to mechanically accessible membranes and has to be checked carefully because of the possibility of precipitated crystals with sharp edges which may cut the membrane while being mechanically moved along the module (80).

Chemical methods

Biocides

The most common approach to biofouling problems in industrial water systems is the use of biocides, or, more correctly, microbicides. Sometimes they are referred to as disinfectants (81), but this might be misleading, because an industrial system will not be disinfected, i.e. cleared from pathogenic microorganisms (82); in the best case, the total number of cells in the water will be drastically reduced. Biocides have been subject to numerous reviews (see 83) and will not be discussed exhaustively in this chapter. However, some of them shall be briefly discussed in a tabular form (Table 2):

58

Table 2. Some biocides in technical use

Chlorine	<u>Advantages:</u> broad spectrum of activity, residual effect, advanced technology available, can be generated on site, active in low concentrations, destroys biofilm matrix and supports detachment <u>Disadvantages:</u> toxic byproducts (84, 85), degradation of recalcitrant compounds to biodegradable products, development of resistance, corrosiveness, reacts with extracellular polymer substances (EPS) in biofilms, low penetration characteristic in biofilms, oxidizes S^{2-} to elemental sulfur (extremely difficult to remove from surfaces (86))
Hypochlorite	<u>Advantages:</u> cheap, effective, destabilizes and detaches the biofilm matrix (25), easy to handle; used for biofilm thickness control (87) <u>Disadvantages:</u> poor stability, oxidizing, rapid aftergrowth observed (88), toxic byproducts, corrosive; does not control initial adhesion (87)
ClO$_2$	<u>Advantages:</u> can be generated on site, activity less pH-dependent, less sensitivity against hydrocarbons, effective in low concentrations <u>Disadvantages:</u> explosive gas, safety problems, toxic byproducts
Chloramine	<u>Advantages:</u> good penetration of biofilms (62-64, 89), reacts specifically with microorganisms; less toxic byproducts, higher residual effect because of lower reactivity with water ingredients (21) <u>Disadvantages:</u> less effective than chlorine to suspended bacteria (21, 90), resistance observed
Bromine	<u>Advantages:</u> very effective against broad microbial spectrum <u>Disadvantages:</u> toxic byproducts, development of resistance
Ozone	<u>Advantages:</u> similar effectivity as chlorine, decomposes to oxygen, no residues, weakens biofilm matrix (Flemming u. Schaule, unpub.) <u>Disadvantages:</u> oxidizes bromide in seawater, reacts with organics and can form epoxides (91), degrades humic acids and makes them bioavailable (92-94), corrosive, short half life time, sensitive to water ingredients
H$_2$O$_2$	<u>Advantages:</u> decomposes to water and oxygen, relatively non-toxic, can easily be generated in situ; weakens biofilm matrix and supports detachment and removal (48) <u>Disadvantages:</u> high concentrations (> 3 %) necessessary, frequent resistance, corrosive (95)

Table 2, continued

Peracetic acid   Advantages: very effective in small concentrations, broad
spectrum, kills spores, decomposes to acetic acid and water
(50, 96), no toxic byproducts known, penetrates biofilms (Flemming
et al., unpubl.)
Disadvantages: corrosive, not very stable, increases DOC

Formaldehyde   Advantages: low costs, broad antimicrobial spectrum, stability, easy
application
Disadvantages: Resistance in some organisms (36), toxicity, suspected
to promote cancer, reacts with protein fixing biofilms on surfaces
(61), legal restrictions

Glutaraldehyde   Advantages: effective in low concentrations, cheap, non-oxidizing,
non corrosive
Disadvantages: does not penetrate biofilms well (91), degrades to
formic acid, raises DOC

Isothiazolones   Advantages: effective at low concentrations, broad antibiotic spectrum
(97)
Disadvantages: problems with compatibility with other water ingre-
dients, inactivation by primary amines

Quarternary ammonia compounds (QUACs)
Advantages: effective in low concentrations, surface activity supports
biofilm detachment, relatively non-toxic, adsorb to surfaces and
prevent biofilm growth (56)
Disadvantages: inactivation by low pH, Ca^{2+}, Mg^{2+} (97), development
of resistance (98)

Cleaners

Cleaners are applied to remove biofilms from surfaces. They represent a vast field
of proprietary knowledge and are formulated from a wide range of chelating agents, surfactants, detergent builders, antideposition aids, enzymes and biocides to provide the user
with effective, safe and easy to use products. (99). The type of cleaner required depends
on the type of system and the nature of the foulant. The best method of foulant identification is by extensive analysis of the foulant gathered from a destructive autopsy of the
deposit. But before that, a first step may be to obtain a thorough chemical and biological
analysis of the current feedwater to see if it has changed from the water the system was
designed to process. Knowledge of the source of the feed water alone often is enough to
make a good guess about the type of fouling that is likely to occur (99). Additionally, it

is very important to check the compatibility of the cleaner with other water treatment chemicals and with the materials used in the system.

Whittaker et al. (53) investigated both the biocidal and the biofilm removal efficiency of various cleaning agents on reverse osmosis membranes. They found that some cleaners killed the biofilm without removing it and vice versa. Removal of the biofilm, however, is much more important for the restoration of the plants performance than killing it and leaving it on the surface. A proprietary cleaning mixture named "Biz" proved to be the most effective, but it had the drawback of its oxidizing properties (53).

It has been proposed that dead cells can be removed easier than living cells (68). Experiments dealing with the removal of primary biofilms from various membrane materials, however, showed clearly that dead cells adhered as strongly as living cells (Flemming & Schaule, in prep.). In order to minimize the effect of biofouling, it is probably best to clean RO membranes at frequent and regular intervals (e.g., a minimum of every 30 days), before an extensive biofilm has had an opportunity to develop (100). Membrane cleaning solutions become noticeably less effective as a function of biofilm age (53, 101, 102).

Especially in high purity water systems, the quality of the rinsing water can be crucial for the success of a cleaning measure The demands to the quality of the cleaning water have been formulated as follows (95): Fe <0,05 ppm, Mn <0,02 ppm, chlorine <0,1 ppm, silica <40 ppm and cell number <1000/ml.

Aftergrowth and the recovery of the biofouling deposit

Numerous investigators and plant operators have observed a disappointingly rapid resumption of biofouling immediately following biocide treatment and have termed this phenomenon "regrowth" or "aftergrowth". Characklis (8) proposes "recovery" as a more appropriate term, since growth may only be one of the processes contributing to reestablishment of the biofilm. Figures 3a and b show a biofouling layer on a reverse osmosis membrane before and after prolonged treatment with the most effective among different cleaning solutions (54, 103). Recovery may be due to one or all of the following (8):

1. The remaining biofilm contains enough viable organisms to preclude any lag phase in biofilm accumulation as observed on clean heat exchanger tubes. Thus, subsequent biofilm accumulation after shock treatment is more rapid than on a clean surface.
2. The residual biofilm imparts an increased relative roughness to the surface and thus enhances transport and sorption of microbial cells and other compounds to the surface. The roughness of the deposits may provide a "stickier" surface.
3. Chlorine for example reacts preferentially with the EPS and removes it, and does not reach the biofilm cells, thus leaving biofilm cells more exposed to the nutrients when chlorination ceases.
4. EPS is rapidly created by surviving organisms as a protective response to irritation by chlorine
5. There is a selection of organisms less susceptible to the biocide which proliferate better between successive biocide applications.

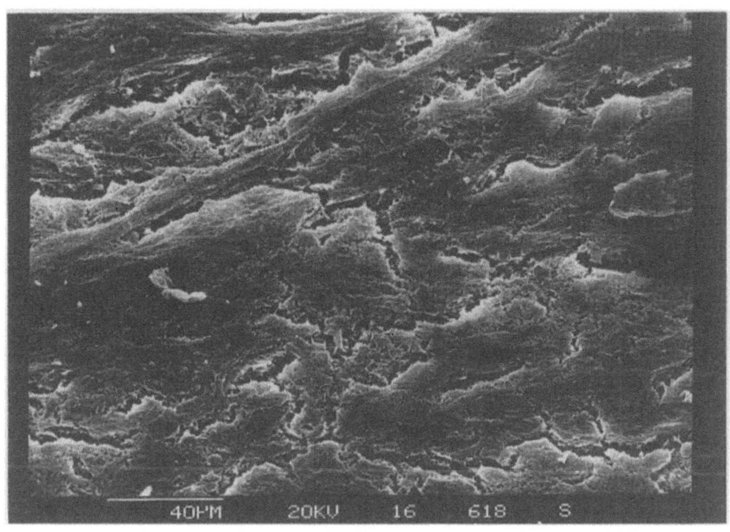

Fig. 3a. SEM of a surface of an irreversibly blocked RO membrane

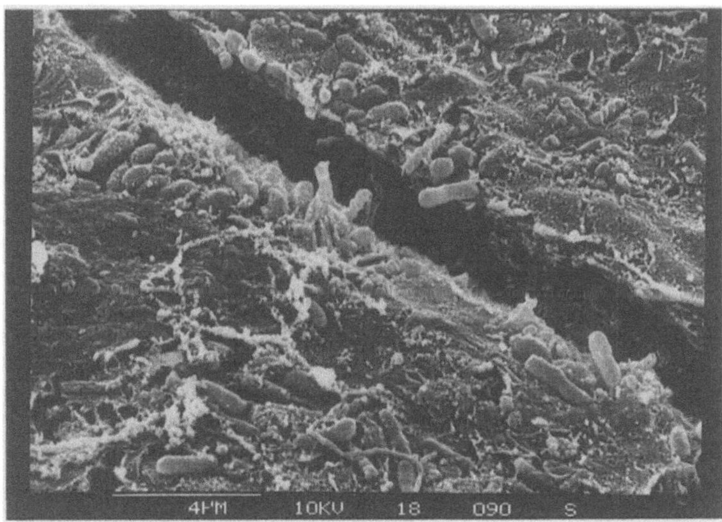

Fig. 3b. Surface of the same membrane after 2 x 24 h treatment with a cleaning agent. 80% of the biofouling deposit has been removed (54)

62

Monitoring the effectivity of sanitization

The problem of the recovery of the biofouling deposit and "aftergrowth" of cells in the system is closely related to how the effectivity of the sanitization is monitored. In almost all practical cases, improvement of plant performance parameters is used to estimate the effectivity of the cleaning process. Flemming and Schaule (54) found that the membrane cleaners applied to a mature biofilm from a membrane surface could not remove more than 80% of it (Fig. 3). Such a reduction of the biofouling layer would lead to a significant improvement of the system. However, the 20% biofilm remaining will enhance the recovery of the biofouling deposit as above discussed, making the effect of the sanitization as transient as it is observed in practice (53). As noted before, cell numbers in the water phase are not indicative for the biofilm situation.

Monitoring of a few representative parts of the system can give the information about the biofilm situation. As biofouling is a process occurring on all surfaces of the system in overall contact with water, inspection of significant parts could be diagnostic. Prefilters are often accessible and can be checked. More methods to monitor biofouling development are discussed later and in chapter 2.

PREVENTION OF BIOFOULING

Up to now, the "silver bullet" against biofouling has not yet been invented. In this context, the prevention of biofouling cannot be achieved by single measurements or "once and forever", but by a strategy based on a *"clean system philosophy"*. The whole system has to be considered, because even if sensitive parts are monitored and cleaned quite thoroughly, other parts may introduce significant amounts of biomass into the system.

Assessment of the Biofouling Potential

The first prerequisite for effective control and prevention of biofouling is: to be aware that a biofouling problem can occur. Many biofouling problems come to maturity only because they have been ignored for a long period of time. Thus, we have to systematically consider the components of a water treatment facility (Table 3).

The fact that stagnant systems frequently contain much higher cell concentrations in the water than flowing systems, led to the recommendation to recirculate the water during off-periods in order to avoid microbial aftergrowth. However, microbial growth is by no way suppressed by the flow; the new cells are only diluted. Experiments with two ion exchanger units - one stagnant, one recirculated - showed clearly that after short time the cell densities in the water became similarly high (70). Thus, recirculation may only result in a fermenter system, operating on a low nutrient level. In stagnant zones there is usually not an enhanced growth but only a local accumulation of cells which are not transported by the flowing system.

Table 3. Biofouling potential

High potential	Low potential
Plant design (see chapter 7)	
extended piping	short piping
dead legs, fissures	no dead legs or fissures
armatures with dead zones	specially designed armatures
access of light	exclusion of light
materials leaching nutrients	inert materials
rough surfaces	smooth surfaces
non disinfected holding tanks	disinfectable holding tanks
Feedwater characteristics	
temperatures > 25 °C	low temperatures
high concentrations of	low concentrations of
- organic nutrients	- organic nutrients
- inorganic nutrients	- inorganic nutrients
- cells	- cells
- particles	- particles
Operational parameters	
intermittent operation	continuous operation
rare monitoring of biofouling	good monitoring of biofouling
unattended ion exchanger and activated carbon filters	limited and monitored filter systems
changing raw water composition	constant and controlled raw water composition
contaminated chemicals	clean chemicals
claning only after failures	early and preventive cleaning
bad access to surfaces	good access to surfaces
insufficient efficiency of cleaning measures	good efficiency control of cleaning measures

The depletion of nutrients may be one of the major factors to keep biofilm accumulation low, although it will be extremely difficult to exclude it in this way completely. The assessment of the assimilable organic carbon (AOC) (8) and the tendency of microbial aftergrowth in a given raw water (104) are very important parameters in this respect.

Thus, organic carbon concentration, flux and biodegradability are important considerations in any attempt to predict the rate and extent of biofouling (8). One has to keep in mind that in nutrient limited systems the increase of shear forces also increases the nutrient load. Thus, higher shear forces can lead to an increase of biofilm accumulation under those circumstances.

It is very important to check the *microbiological quality of the pretreatment chemicals* used for purposes other than the prevention of biofouling. Sodium hexametaphosphate (SHMP), used as a scaling inhibitor, has been reported to contain high numbers of bacteria and to provide orthophosphate as inorganic nutrient for the fouling biofilm (86, 105). Thus, SHMP had to be treated with sodium metabisulphite (SMBS) for sanitization. SMBS is used for dechlorination prior to RO membranes. It has biostatical properties which are utilized either by shock dosage (ca. 500 mg/l) or continuous dosage (20-50 mg/l) in practical applications.

Monitoring of Biofouling

Water quality monitoring

A regular water analysis, checking salt content, COD, pH and conductivity normally gives no clues to predict the biofouling potential. If biofouling is expected, the microbial content of the water is checked, usually with cultivation methods. George et al. (106) investigated the microbial content of three water samples, drawn at three different sites of an RO plant. The numbers gained with cultivation methods were always the lowest compared to microscopical methods, indicating the cells which are able to grow under the given laboratory conditions. However, epifluorescence microscopy (EM), not discriminating between living and dead cells, indicates the *maximal number of cells* present in the water sample. INT (iodonitrotetrazolium violet) is a fluorescent dye activated by microbial activity, that distinguishes active from non-active cells, when examined by direct epifluorescence microscopy. Diacetylfluorescein (107) is deacetylated by the hydrolases of living bacteria and can be used for viable microorganisms detection. There is no reliable correlation between cultivation and the microscopical methods. However, it is recommended to monitor the bacteriological quality of the raw water in order to assess the biofouling potential, because the raw water is the main source for the biofilm microorganisms. The following techniques are most commonly included in a water biomonitoring program:

* Acridine orange direct count
* INT (living cell number) direct microscopic technique
* Standard serial-dilution plating
* Membrane filtration
* Anaerobe enumeration with API broth
* Pyrogen measurement (108)

Details about the methods are given by Applegate and Erkenbrecher (109) and (including biofilm sampling and analysis) Geesey and Mittelman (110).

Surface monitoring

Reliable information about the biofouling situation in a water treatment system is highly desirable in order to prepare for future problems, intervention and evaluation of the effectivity. If a biofilm is visible with the naked eye, the system is already in an advanced phase of biofouling. Generally, two principles of monitoring can be used:

a) On-line fouling monitors in the water system
b) Side-stream fouling monitors

The choice is dependent upon operational and economical conditions of a given unit. The following considerations should be incorporated in a monitoring program (91):

1. Water quality in the system (establish the basis for flushing frequencies and wet layup treatments)
2. Design a facility for testing the impact of proposed water treatment programs on various system components and water system materials
3. Determine biofouling/biocorrosion rates in water system under stagnant, low flow, and high flow conditions. Address both short and long term rates for general and localized pitting corrosion
4. Monitor in various valves, and localized galvanic corrosion at welds
5. Provide for visual inspection and removal of samples for destructive analysis
6. Provide for expansion of sidestream facility to include additional fouling/corrosion testing equipment

When biofilms are visually detectable, the accumulation process has already progressed to a late phase. It is a common observation and well known that biofilms are the more difficult to remove the older they are. Thus, monitoring systems indicating the early stage are very desirable. Generally, direct or indirect methods are available.

Direct methods address the development of the biofilm on a given surface and indicate the presence of cells. Indirect methods respond to the effect of the biofilm - they may be physical (drag resistance, heat transfer, mass transfer, light absorption) or chemical (pH-shifts, enzyme activity, degradation, production of metabolites).

Effects like changes in drag resistance, heat transfer resistance, or mass transfer resistance only occur when the thickness of the biofilm exceeds the thickness of the viscous sublayer in the flowing system (91). This means that most of the indirect methods do not address early biofilm formation. However, in systems where biofouling has to be kept below a certain level, they may serve as valuable tools.

Some monitoring devices

The "Robbins Device" provides another tool for monitoring biofilm development (111; details: see 110). Principally, it consists of a removable part of the water systems surface which can be microbiologically examined after certain times. This system helps to assess the general biofouling situation but it is not suitable to detect it at all critical places.

Other - indirect - biofilm monitoring systems (BFM) can be adapted from different applications such as biofouling control in heat exchangers. They are based on measuring the increase in drag resistance (112, 113). Pressure drop in the BFM - and hence the friction factor - can be correlated with development of biofilm in the system. Characklis et al. (114) described a monitoring device which continuously monitors the deposition of microbial biomass on surfaces indirectly by measuring changes in heat transfer and fluid frictional resistance. The microcomputer-controlled system simulates conditions by maintaining the appropriate process fluid velocity (or pressure drop) and heat flux (or wall temperature) to any metal surface on which deposition is occurring. Howsam & Tyrell (115) proposed a simple device on the basis of a mini sand filter to monitor biofouling in groundwater systems.

Reverse osmosis technique as an example of biofouling monitoring

Reverse osmosis technology may serve as an example of a technique which is very sensitive to biofouling. In most cases, biofouling is recognized by indirect effects on the unit's performance: flux decline, decrease in salt rejection or an increase in the delta-p feed/brine. However, these are non-specific reactions of the system and without further information, cannot easily be correlated to biofouling, as indicated in phase I and II of Fig. 1.

Therefore, it is often impossible to conclude that a certain increment of flux decline is the result of a certain amount of microbial fouling on the membrane surface. Ridgway (9) reports, that within the first three days of operation, approximately 2×10^5 bacteria per cm^2 (measured as total CFU) were recovered from the surfaces of the cellulose acetate membranes, although no visible evidence of biofouling could be observed by the unaided eye. Within one or two weeks of initial operation, the fouling bacteria produced a nearly confluent lawn (approximately one cell layer in thickness) extending over the entire membrane surface (9). Flemming et al. (in prep.) observed this effect even after a few days. Thus, biofilm formation occurs before plant performance parameters respond. Therefore, monitoring should be carried out as directly as possible.

A reasonable approach to control RO-biofouling is the use of "sacrificial elements" as biofouling monitoring systems. Winters and Isquith (116) and Ridgway et al. (1985) used mini RO probes which were integrated on line in the system and removed from time to time to check biofouling directly. Winters and Isquith (116) proposed an easy colorimetric test, based on the specific reaction of the dye Alcian Blue with extracellular acidic polysaccharides (117). Practical experience with this approach have not been reported.

The ideal monitoring system

The requirements of an ideal biofouling monitoring system include:

* Monitoring of biofilm formation on representant surfaces
* Monitoring of the microbial quality of feedwater and all added pretreatment chemicals
* Monitoring of plant performance data in order to evaluate correlations which allow one to predict and recognize biofouling, and allow one to establish the individual fouling characteristics of an RO plant.

All this is monitored in real time, on line, non destructively and automatically and costs almost nothing. Unfortunately there is no such system available yet. Some new monitoring techniques such as the automatic image analyzer, the FTIR technique, the Quartz crystal microbalance (118) or the electrical impedance spectroscopy may eventually be incorporated into routine system monitoring programs (see chapter 7).

Surface coatings and modifications

Schoenen (52) has summarized his extensive work on coatings of materials in contact with drinking water (39, 40): materials organic based materials (coatings, films, sealants, plastic pipes and hoses) or with organic additives (cement mortar with organic compounds) can promote an intensive growth of microorganisms. Field observations with bituminous and epoxy resin coatings, PVC films, polyamide pipes and a plastic-containing cement show, that there is a considerable **increase** of growth of microorganisms in the water, compared to non-leaching materials, and a visible microbial growth upon the surface of the materials. He correlates this with the release of biodegradable matter from the substratum, observing much less microbial growth on "inert" materials. However, his work was focused on the cell density in the water phase. O'Connor & Bannerji (31), inspecting the biofilm accumulation on the surface of various materials, could not observe any significant differences between the colonization of PVC, copper or iron surfaces. Thus, biofilms on different materials may release cells in different amounts to the bulk liquid phase, when they can leach nutrients from the support.

It is an aim of biofouling control research to find a way to make the surfaces more "hostile" to microorganisms. An extreme example is the development of carrier-bound disinfectants (119), which has been reported to be successful even in killing suspended microorganisms. Experiments with radioactively-marked, covalently-bound disinfectants showed that the effect is not due to released material. However, some basic questions are unresolved, apart from the mechanism of action - such as how long the biocidal efficiency is maintained when organic molecules are adsorbed and how many cells stay attached to the surface after they have been killed.

A more common application of "surface hostility" may be the utilization of the "oligodynamic effect" of silver ions, which has been known since the end of the last century (120) and exploited for water treatment (121) in the so-called Katadyn process. Metallic silver releases silver ions in very low concentrations to the water phase where they exert a biocidal effectivity on the microbes. We investigated this effect on ion exchangers, which were partially silver-coated to prevent microbial aftergrowth (122, 123). At first, the bactericidal effect of silver ions could be reproduced very well, but after a period of some weeks, microbial growth was observed in the presence of "oligodynamic" silver concentrations, i.e. 50-100 μg/l of Ag$^+$. The term "oligodynamic" was coined to describe biological effects at very low ("oligo") concentrations. In the dimensions of weeks, silver coating did not prevent microbial adhesion and growth. The microflora was able to adapt to the presence of silver, as shown in Fig. 4.

A silver-tolerant population had developed on a silver-coated ion exchanger. The silver tolerance was checked in a simple test (124) and expressed as toxic limit concentration (TLC). The disinfection with 0,01% peracetic acid extinguished the silver tolerant microflora completely, as shown by the germ-effectivity of the silver, proven by the low TLC in the subsequent period. However, a gradual increase of cell number was observed, when the TLC was rising. At the end of the experiment, the TLC was far above the silver concentration present in the water. We could reproduce the rise of silver tolerance in tap water microbial population by the selection pressure of sublethal silver concentration. Additionally, we investigated the silver tolerance of the original tap water microflora and found that it varied randomly between 5 and 75 μg/l Ag$^+$ (123). This explained to us why silver coating did not work in the long term in most practical applications.

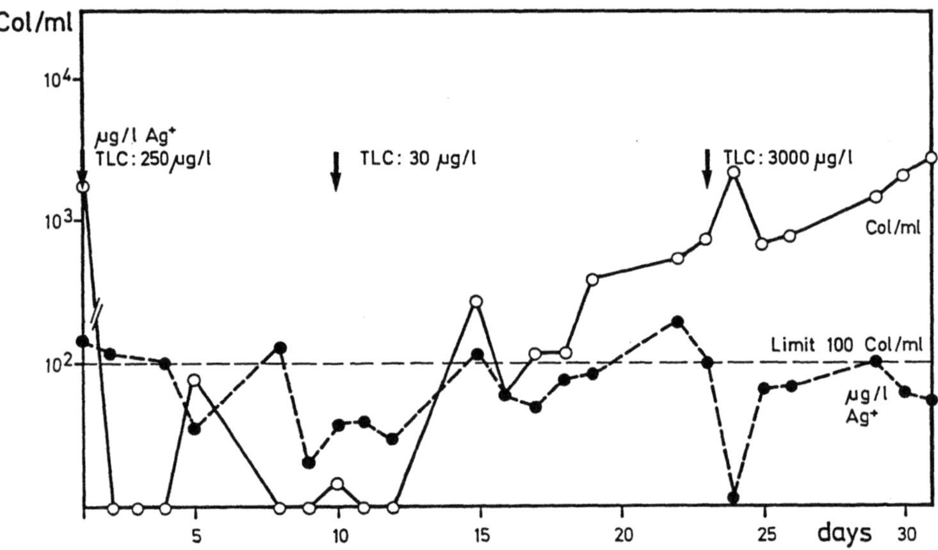

Fig. 4. Microbial growth in a silver-coated ion exchanger unit.
o—o = cell numbers per ml wet ion exchanger resin, ●---● = silver concentration in μg/l (122, 123)

The above considerations support various statements (1, 71) that there are no practical examples of industrial biofouling control achieved solely by the application of surface coatings. For the prevention of biofouling, continuous efforts and a rational strategy are necessary.

DO WE HAVE TO LIVE WITH BIOFOULING?

In most cases, it will be too difficult to prevent the accumulation of biofilms. We have to consider that there is no material in the environment which cannot be colonized by microorganisms, given the right conditions - and the right conditions are present in practically all compartments of a water treatment facility. The example of ultrapure water production in microelectronics shows extensive requirements for effectively preventing biofilm growth; such efforts are not acceptable for the majority of industrial processes.

Thus, biofilms can grow industrial systems. When they start turning into a biofouling problem, biofilm accumulation has reached the plateau phase already and the processes of initial adhesion long since been completed. The question is now: how can the level of biofilm accumulation be kept below that treshold?

In this context it is interesting to consider what governs the extent of biofouling accumulation, i.e. the level of the plateau. The changes in mass balance in the steady state phase can be described in a very simple equation:

$$\Delta M_{adhesion} + \Delta M_{growth} = \Delta M_{detachment}$$

The common response to a biofouling problem is the disinfection of the raw water. This addresses the cells in suspension and thus, $\Delta M_{adhesion}$, which is formed by suspended cells adhering to the biofilm surface. However, Bott (125) showed in experiments on biofouling of heat exchangers, that the thickness of a given biofilm was not influenced when the supply of bacteria in the bulk liquid was shut off. As soon as the nutrient level of the bulk liquid was lowered, the biofilm thickness decreased in response. This corresponds to the observations of Characklis (127) that the adhesion of new cells from the bulk liquid is not the factor which contributes significantly to biofilm accumulation in the plateau phase.

ΔM_{growth} is much more important, although the nutrient concentration in the water is more difficult to control than the concentration of suspended cells. Cutting down the nutrient concentration certainly will not prevent biofilm growth in general. However, it will help to decrease biofilm thickness. In a nutrient limited system, the use of oxidizing biocides can increase biofilm growth by degrading recalcitrant substances into biodegradable substrates (89).

Another common answer in response to biofouling is flushing the system rigorously, followed by various techniques such as air bumping, addition of cleaners or pigging. This addresses $\Delta M_{detachment}$, which is the result of the balance between the shear forces and the mechanical stability of the biofilm matrix. In this case, the mechanical properties of the deposit become significant (48), which may be controlled by

a) Factors which can stabilize the biofilm, such as

- Nature of the EPS (can control gel stability)
- Filaments and fibers in the matrix (improve internal stability of the matrix)
- Roughness of the substratum (improves adhesion to the support)
- Nutrient situation (low nutrient levels increase matrix stability)

b) Factors which can destabilize the biofilm, such as

- Oxidants
- Solvents
- Gas bubbles (e.g. from anaerobic processes in the base biofilm)
- Complexation of Calcium ions (127)
- shocks of pH, ionic strength, temperature
- Gradients in pH, reaction products, other dissolved substances
- Lysis of cells (can destabilize the matrix)
- Roughness of the biofilm surface (increases shear forces the biofilm is exposed to)
- Mechanical stress, caused by
    * shear forces
    * ultrasonic energy (77, 78)
    * temperature stress (supercooled liquids [74])
- Smoothness of the support (decreases the adhesion force of the biofilm to the substratum)

c) Factors which keep the extent of biofouling deposit accumulation low, such as

- low nutrient concentration and load
- strong shear forces, changing shear forces
- presence of biocides such as chlorine, ozone etc.
- low particle concentrations

It is clear that this list is incomplete and that there is a strong demand for more research in that area.

For a given system, these factors have to be considered in order to design a strategy which allows one to keep the accumulation of the biofilm below a level of interference with the production demands. It is obvious, that this can differ from system to system drastically, depending on the actual conditions.

A "clean system philosophy" must include the observation of the surfaces. It has to be combined with the awareness of the possibility of biofouling problems. Some - possibly very simple - operational precautions can curb biofilm growth and provide protection. As a general rule, early diagnosis and recognition of biofouling allows early reaction and prevention of strong biofouling development.

Learning from natural systems

With this background it may be interesting to find out how some marine animals manage to stay free from microbial colonization in an open, non-sterile system (128), where practically every surface is covered by microorganisms. Vrolijk et al. (129) investigated the fact that certain gorgonian octocorals, are naturally fouling-resistant. They measured a low surface energy (23-27 mN/m) in the range of "minimal adhesiveness". They state that biofouling is a complex phenomenon and that it is improbable that only one strategy could effectively inhibit the wide range of potential fouling organisms. They suggest that "minimally adhesive" surfaces based on low energy outermost layers may be a primary, passive antifouling defence in gorgonian corals. *Polysyncraton lacazei*, a marine organisms, defends its surface by shedding its skin, excretion of antimicrobial, antifungal, antialgal and antimitotic agents and of proteolytic enzymes which block the movement of cilia of colonizing larvae, and as a "polish", there are small crustaceae (Harpaticoids) and mites, which graze algae and fungi which broke through the defense mechanisms (130).

ACKNOWLEDGEMENTS

The financial support of the Willy-Hager-Stiftung to this work is gratefully acknowledged. The excellent technical assistance of Hanna Rentschler and the work of Gabriela Schaule have provided a substantial basis for which I am most thankful.

REFERENCES

1   Lappin-Scott HM, Costerton JW (1989) Bacterial biofilms and surface fouling. Biofouling 1, 323-342

2   Sternel C (1989) Feinere Chipstrukturen erfordern extrem keim- und partikelarme Anlagen. Reinraumtechnik 2/89, 16-20

3   Larsson E (1984) Betriebserfahrungen mit einer Anlage zur Erzeugung von Wasser für pharmazeutische Zwecke mittels des Umkehrosmose-Verfahrens. In: Dlouhy G, Marquardt K (eds) Moderne Techniken zur Wasseraufbereitung für pharmazeutische, kosmetische und medizintechnische Zwecke. Hager + Elsässer, Ruppmannstr. 22, D-7000 Stuttgart

4   White DC, Mittelman MC (1990) Biological fouling of high purity waters: mechanisms and consequences of bacterial growth and replication. Semicond. Pure Water Conf., Santa Clara, CA, Jan. 17-18; 150-171

5   Morita RY (1985) Starvation and miniaturisation of heterotrophs with special emphasis on maintenance of the starved viable state. In: M. Fletcher u. G. Floodgate (Hrsg.) Bacteria in their natural environments: the effect of nutrient conditions. Soc. Gen. Microbiol., U.K., 1985; 111-130

6   Kjelleberg S (1985) Mechanisms of bacterial adhesion at gas-liquid interfaces. In: Savage DC, Fletcher MM (eds)(1985) Bacterial adhesion. Plenum Press New York London; 163-194

7   Schlegel HG (1985) Allgemeine Mikrobiologie. Thieme, Stuttgart

8   Characklis WG (1990) Microbial fouling. In: Characklis WG, Marshall KC (eds) Biofilms. John Wiley, New York, 523-584

9   Ridgway HF (1988) Microbial adhesion and biofouling of reverse osmosis membranes. In: Parekh BS (ed) Reverse osmosis technology. Marcel Dekker, New York, Basel; 429-481

10   Cunningham AB, Bouwer EJ, Characklis WG (1990) Biofilms in porous media. In: Characklis WG, Marshall KC (eds) Biofilms. John Wiley, New York; 697-732

11   Flemming HC (1987) Microbial growth on ion exchangers. Wat. Res. 21, 745-756

12   Scheer R (1988) MIcrobial growth in mixed-bed ion exchange columns under differing conditions. Zbl. Bakt. Hyg. B 185, 379-396

13   Edyvean RGJ (1990) Fouling and corrosion in water filtration and transportation systems. In: Howsam P (ed) Microbiology in civil engineering. Chapman and Hall, London; 62-74

14   Camper AK, LeChevallier MW, Broadway SC, McFeters GA (1986) Bacteria associated with granular activated carbon particles in drinking water. Appl. Environ. Microbiol. 52, 434-438

15   Dott W, Rollinger Y (1987) Survival and interactions of selected bacterial species in activated carbon filters. Zbl. Bakt. Hyg. B 184, 525-537

16   Fiore JV, Babineau RA (1977) Effects of an activated carbon filter on the microbial quality of water. Appl. Environ. Microbiol. 34, 541-546

17   Geldreich EE, Taylor RH, Blannon JC, Reasoner DJ (1985) Bacterial colonization of point-of-use water treatment devices. J. Am. Water Works Assoc. 87, 72-80

18   Geller A (1983) Growth of bacteria in inorganic medium at different levels of airborne organic substances. Appl. Environ. Microbiol. 46, 1258-1262

19   Seiferth R, Krüger W (1950) Überraschend hohe Reibungsziffer einer Fernwasserleitung. VDI-Zeitschr. 92, 189-191

20   Characklis WG (1973) Attached microbial growths - II. Frictional resistance due to microbial slimes. Wat. Res. 7, 1249-1258

21   v.d.Wende E, Characklis WG (1990) Biofilms in potable water distribution systems. In: McFeters GA (ed) Drinking water microbiology. Springer International, New York, 249-268

22   Mittelman MW, Nivens DE, Low C, White DC (1990) Differential adhesion, activity, and carbohydrate:protein ratios of *P. atlantica* monocultures attaching to stainless steel in a linear shear gradient. Microb. Ecol. 19, 269-278

23   Vogel F, Exner M, Tuschewitzki GJ, Leinhos C (1984) Adhäsion von Mikroorganismen an Beatmungstuben. Dt. Med. Wochenschr. 109, 1148-1150

24   Jacques M, Marrie TJ, Costerton JW (1987) Review: microbial colonization of prostethic devices. Microb. Ecol. 13, 173-191

25   Costerton JW, Cheng KJ, Geesey GG, Ladd TJ, Nickel JC, Dasgupta M, Marrie TJ (1987) Bacterial biofilms in nature and disease. Ann. Rev. Microbiol. 41, 435-464

26   Schoenen D (1985): TOC, TIC und TOC-Bestimmung an Werkstoffen für den Trinkwasserbereich. Z. Wasser Abwasser Forsch. 18, 254-257

27   LeChevallier MW, Babcock TM, Lee RG (1987) Examination and characterization of distribution system biofilms. Appl. Environ. Microbiol. 53, 2714-2724

28   Ridgway HF, Olson BH (1981) Scanning electron microscope evidence for bacterial colonization of a drinking water distribution system. Appl. Environ. Microbiol. 41, 274-287

29   Allen MJ, Taylor RH, Geldreich EE (1980) The occurrence of microorganisms in water main encrustations. J. Am. Water Works Assoc. 82, 614-625

30   Tuovinen OH, Hsu JC (1982) Aerobic and anaerobic microorganisms in tubercles of the Columbus, Ohio, water distribution system. Appl. Environ. Microbiol. 44, 761-764

31   O'Connor JT, Banerji SK (1984) Biologically mediated corrosion and water quality deterioration in distribution systems. EPA-600/52-84-056. U.S. Environmental Protecting Agency, Cincinnati, OH

32   v.d.Wende E, Characklis WG, Grochowski J (1988) Bacterial growth in water distribution systems. Wat. Sci. Tech. 20, 521-524

33   v.d.Wende E, Characklis WG, Smith DB (1989) Biofilms and bacterial drinking water quality. Wat. Res. 23, 1313-1322

34   Carson LA, Bland LA, Cusick LB, Favero MS, Bolan GA, Reingold AL, Good RC (1988) Prevalence of nontuberculous mycobacteria in water supplies of hemodialysis centers. Appl. Environ. Microbiol. 54, 3122-3125

35   Schulze-Röbbecke R, Fischeder R (1989) Mycobacteria in biofilms. Zbl. Bakt. Hyg. 188, 385-390

36   duMoulin GC, Stottmeier KD, Pelletier PA, Tsang AY, Hedley-White J (1988) Concentration of *Mycobacterium avium* by hospital hot water systems. J. Am. Med. Assoc. 260, 1599-1601

37   Schoenen D, Dott W (1977) Microbial growth on an expansion joint of a reservoir.
     Zbl. Bakt. Hyg. Orig. B 165, 464-470

38   Dott W, Schoenen D (1985) Qualitative and quantitative examination of bacteria
     found in aquatic habitats. 7th communication: development of bacterial aufwuchs
     on different working materials exposed to potable water. Zbl. Bakt. Hyg. Orig. B
     180, 436-447

39   Schoenen D, Schöler HF (1985) Microbial settlement of paint- and building materi-
     als in the sphere of drinking water. 11th communication: long term inspection of
     the polyester isolation of a drinking water reservoir. Zbl. Bakt. Hyg. 180, 429-435

40   Schoenen D, Schulze-Röbbecke R, Schirdewahn N (1988) Microbial contamination
     of water by materials of pipes and hoses. 2nd communication: Growth of *legionella
     pneumophila*. Zbl. Bakt. Hyg. 186, 326-332

41   Tuschewitzki GJ, Exner M, Thofern E (1983) Induction of microbial growth in flexi-
     ble plastic tubes by drinking water. Zbl. Bakt. Hyg. B 178, 380-388

42   Exner M., Tuschewitzki GJ, Thofern E (1983) Observations of bacterial growth on
     a copper pipe line of a central disinfection dosage apparatus. Zbl. Bakt. Hyg. Orig.
     B 177, 170-181

43   McIntyre F (1974) The top millimeter of the ocean. Scient. Am. 230, 62-77

44   Kjelleberg S (1985) Mechanisms of bacterial adhesion at gas-liquid interfaces. In:
     Savage DC, Fletcher, MM (eds.) Bacterial adhesion. Plenum Press, New York; 163-
     194

45   Hermansson M, Kjelleberg S, Korhonen TK, Stenström TA (1982) Hydrophobic
     and electrostatic characterization of surface structures of bacteria and its relation-
     ship to adhesion to an air-water interface. Arch. Microbiol. 131, 308-312

46   Kefford B, Kjelleberg S, Marshall KC (1982) Bacterial scavenging: utilization of
     fatty acids localized at a solid-liquid interface. Arch. Microbiol. 133, 257-260

47   Thofern E, Botzenhart K (1974) Untersuchungen zur Verkeimung von Trinkwasser.
     Gas Wasser Abwasser 115, 459-460

48   Christensen BE, Characklis WG (1990) Physical and chemical properties of biofilms.
     In: Characklis GW, Marshall KC (eds) Biofilms. John Wiley, New York; 93-130

49   Flemming HC, Schaule G (1988) Biofouling on membranes - a microbiological ap-
     proach. Desalination 70, 95-119

50   Flemming HC (1984a) Peracetic acid as a disinfectant - a review. Zbl. Bakt. Hyg., I. Abt. Orig. B 179, 97-111 (English translation available from the author)

51   Fletcher MM (1980) The question of passive versus active attachment mechanisms in nonspecific bacterial adhesion. In: Berkeley RCW, Lynch JM, Melling J, Rutter PP, Vincent B (eds) Microbial adhesion to surfaces. Ellis Horwood, Chichester; 197-210

52   Schoenen D (1990) Influence of materials on microbiological colonization of drinking water. In: Howsam P (ed) Microbiology in civil engineering. Chapman and Hall, London, 121-145

53   Whittaker C, Ridgway H, Olson BH (1984) Evaluation of cleaning strategies for removal of biofilms from reverse-osmosis membranes. Appl. Environ. Microbiol. 48, 395-403

54   Flemming HC, Schaule G (1989) Investigations of biofouling on reverse osmosis and ultrafiltration membranes. Part II: Analysis and removal of surface film. Vom Wasser 73, 287-301

55   Costerton JW, Jones PA (1986) The importance of sessile sampling in the monitoring of bacterial corrosion problems. Can. Reg. West. Conf., Nat. Ass. Corr. Eng., Feb. 25-27, Calgary, Canada

56   Ridgway HF, Rigby MG, Argo DG (1985) Bacterial adhesion and fouling of reverse osmosis membranes. J. Am. Water Works Assoc. 77, 97-106

57   Nichols WW (1989) Susceptibility of biofilms to toxic compounds. In: Characklis WG, Wilderer PA (eds) Structure and function of biofilms. John Wiley, New York, 321-331

58   Costerton JW, Lashen ES (1984) Influence of biofilm on efficacy of biocides on corrosion-causing bacteria. Mat. Perf. 23, 13-17

59   Clark TF (1984) Chlorine tolerant bacteria in a water distribution system. J. Am. Water Works Assoc. 76, 65-67

60   Costerton JW (1984) The formation of biocide-resistant biofilms in industrial, natural, and medical systems. Dev. Ind. Microbiol. 25, 363-372

61   Exner M, Tuschewitzki GJ, Scharnagel J (1987) Influence of biofilms by chemical disinfectants and mechanical cleaning. Zbl. Bakt. Hyg. B 183, 549-563

62   LeChevallier MW, Cawthon CD, Lee RG (1988) Factors promoting survival of bacteria in chlorinated water supplies. Appl. Environ. Microbiol. 54, 649-654

63   LeChevallier MW, Cawthon CD, Lee RG (1988a) Inactivation of biofilm bacteria. Appl. Environ. Microbiol. 54, 2492-2499

64   LeChevallier MW, Lowry CD, Lee RG (1990) Disinfecting biofilms in a model distribution system. J. Am. Water Works Assoc. 92, 87-99

65   Herson DS, McGonigle B, Payer MA, Baker KE (1987) Attachment as a factor in the protection of Enterobacter cloacae from chlorination. Appl. Environ. Microbiol. 53, 1178-1180

66   Costerton JW, Boivin J (1990) The role of biofilm bacteria in microbial corrosion. Int. Congr. Microb. Induc. Corr. October 7-12, Knoxville, TN.

67   Payment P (1989) Bacterial colonization of domestic reverse-osmosis water filtration units. Can. J. Microbiol. 35, 1065-1067

68   Argo D, Ridgway HF (1982) Biological fouling of reverse osmosis membranes. Aqua 6, 481-491

69   Characklis WG, Turakhia MH, Zelver N (1990) Transport and interfacial transfer phenomena. In: Characklis WG, Marshall KC (eds) Biofilms. John Wiley, New York, 265-340

70   Flemming HC (1988) Microbes on ion exchangers. In: Streat M (ed) Ion exchange for industry. Ellis Horwood, Chichester; 82-90

71   McCoy WF (1987) Strategies for the treatment of biological fouling. In: Mittelman MW, Geesey GG (eds) Biological fouling of industrial water systems. Water Micro Associates, San Diego; 247-268

72   Bartz JC, Bell RJ, Mussalli YG, Pope DH, Rosen MD (1987) Results of steam condenser fouling and corrosion studies. Paper pres. at Am. Soc. Mech. Eng. Ann. Winter Meet., Corresp.: Bartz JA, Electric Power Res. Inst., PO Box 10412, Palo Alto, CA, 94303

73   Nickels JS, Bobbie RJ, Lott DF, Martz RF, Benson PH, White DC (1981) Effect of manual brush cleaning on biomass and community structure of microfouling film formed on aluminium and titanium surfaces exposed to rapidly flowing seawater. Appl. Environ. Microbiol. 41, 1442-1453

74   Costerton JW (1983) Biofilm removal. U.S. Pat. 4.419.248

75   Meltzer TH (1987) Ultraviolet and ozone system. In: Mittelman MC, Geesey GG (eds) Biological fouling of industrial water systems - a problem solving approach. Water Micro Assoc., San Diego, CA; 97-137

76   Kreft P, Scheible OK, Venosa A (1986) Hydraulic studies and cleaning evaluations of ultraviolet disinfection units. J. Water Poll. Contr. Fed. 58, 1129-1137

77   Hill FK, Pandolfini PP, Duger GL, Avery WH (1981) Biofouling removal by ultrasonic radiation. Corrosion 81, Int. Corr. Forum, April 6-10, 1981, Toronto; Paper No. 267,

78   Zips A, Schaule G, Flemming HC (1990) Ultrasound as a mean for detachment of biofilms. Biofouling 2, 323-333

79   Yanagi C, Mori K (1980) Advanced reverse osmosis process with automatic sponge ball cleaning for the reclamation of municipal sewage. Desalination 32, 391-398

80   Belfort G (1977) Pretreatment and cleaning of hyperfiltration (reverse osmosis) membranes in municipal wastewater renovation. Desalination 71, 285-300

81   Noss CI, Isaak RA (1990) Disinfection. Res. J. Water Poll. Contr. Fed. 62, 435-440

82   Wallhäußer KH (1988) Praxis der Sterilisation - Desinfektion - Konservierung. Thieme, Stuttgart

83   Payne KR (1988) Industrial biocides. John Wiley, New York

84   McGuire MJ (1989) Preparing for the disinfection by-products rule: a water industry status report. J. Am. Water Works Assoc. 81, 35-66

85   Krasner SW, McGuire MJ, Jacangelo JG, Patania NL, Reagan KM, Aieta EM (1989) The occurrence of disinfection by-products in US drinking water. J. Am. Water Works Assoc. 81, 41-53

86   Ahmed SP, Alansari MS (1989) Biological fouling and control at RAS Abu Jarjur RO plant - a new approach. Desalination 74, 69-84

87   Hassan RS, Oh LCP (1989) Effect of sodium hypochlorite (clorox) and its mode of application on biofilm development. Biofouling 1, 353-361

88   Miller PC, Bott TR (1982) Effects of biocide and nutrient availability on microbial contamination of surfaces in cooling water systems. J. Chem. Tech. Biotechnol. 32, 538-546

89   Applegate LE, Erkenbrecher CW, Winters H (1989) New chloramine process to control aftergrowth and biofouling in Permasep B-10 RO surface seawater plants. Desalination 74, 51-67

90   Wolfe RL, Ward NR, Olson BH (1985) Inactivation of heterotrophic bacterial populations in finished drinking water by chlorine and chloramines. Water Res. 19, 1393-1404

91   Characklis WG (1990) Microbial fouling control. In: Characklis WG, Marshall KC (eds) Biofilms. John Wiley, New York, 585-633

92   Gilbert E (1988) Biodegradability of ozonation products as a function of COD and DOC elimination by the example of humic acids. Water Res. 22, 123-126

93   Jaeggi NE, Schmidt-Lorenz W (1988) Bacterial regrowth in drinking water. 1st communication: drinking water treatment. Zbl. Bakt. Hyg. B 186, 311-325

94   Jaeggi NE, Schmidt-Lorenz W (1988) Bacterial regrowth in drinking water. 2nd communication: distribution system. Zbl. Bakt. Hyg. B 1986, 494-503

95   Lintner K, Bragulla S (1988) Reinigung und Desinfektion von Membrananlagen. Henkel Referate 24, 42-45

96   Baldry MGC, Fraser JAL (1988) Disinfection with peroxygens. In: Payne KR (ed) Industrial biocides. John Wiley, New York; 91-116

97   Hampson JD (1982) Industrial perservatives and biocides. Lab. Pract. 31, 337-338

98   Ventullo RM, Larson RJ (1986) Adaptation of aquatic microbial communities to quarternary ammonium compounds. Appl. Environ. Microb. 51, 356-361

99   Graham SI, Reitz, RL, Hickman CE (1989) Improving reverse osmosis performance through periodic cleaning. Desalination 74, 113-124

100   Trägardh G (1989) Membrane cleaning. Desalination 71, 325-335

101   Ridgway HF (1987) Biological fouling of reverse osmosis membranes: genesis and control. In: Mittelman MM, Geesey GG (eds) Biological fouling of industrial water systems: a problem solving approach. Water Micro Associates, San Diego; 138-193

102   Flemming HC, Schaule G (1988) Investigations on biofouling of reverse osmosis and ultrafiltration membranes. Part I: Initial phase of biofouling. Vom Wasser 71, 207-223

103   Flemming HC (1990) Mechanistic aspects of Reverse Osmosis membrane biofouling. In: Amjad Z (ed) Reverse osmosis technology. Vol I, Van Nostrand Reinhold, New York; in press

104   Werner P (1988) Measurement of the growth rate of bacteria in surface water treatment plants. Vom Wasser 70, 93-105

105   Himelstein WD, Amjad J (1985) The role of water analysis, scale control and cleaning agents in reverse osmosis. Ultapure Water March/April, 32-36

106   George DM, duMoulin G, Carney EM (1988) Comparative evaluation of a renal dialysis ultrapure water system using epifluorescence microscopy, R2A media, Millipore SPC samplers and LAL assay. Abstr. Ann. Meet. ASM L 15

107   Snyder AP, Greenberg DB (1984) Viable microorganism detection by induced fluorescence. Biotechnol. Bioeng. 26, 1395-1397

108   Novitsky T (1987) Bacterial endotoxins (pyrogens) in purified waters. In: Mittelman MC, Geesey GG (eds) Biological fouling of industrial water systems - a problem solving approach. Water Micro Associates, San Diego, CA; 77-96

109   Applegate LE, Erkenbrecher CW (1987) Monitoring and control of biological activity in Permasep seawater RO plants. Desalination 65, 331-359

110   Mittelman MC, Geesey GG (eds)(1987) Biological fouling of industrial water systems - a problem solving approach. Water Micro Associates, San Diego, CA

111   Ruseska I, Robbins J, Lashen ES, Costerton JW (1982) Biocide testing against corrosion-causing oilfield bacteria helps control plugging. Oil Gas J. 253-264

112   Johnson C, Howells M (1981) Biofouling: new insights, new weapon. Power, April 1981, 90-91

113   Donlan RM, Pipes WO (1988) Selected drinking water characteristics and attached microbial population density. J. Am. Water Works Assoc. 80, 70-76

114   Characklis WG, Zelver N, Roe FL (1986) Continuous on-line monitoring of microbial deposition on surfaces. Biodeterioration VI, 427-433

115   Howsam P, Tyrrel S (1989) Diagnosis and monitoring of biofouling in enclosed flow systems - experience in groundwater systems. Biofouling 1, 343-351

116   Winters H, Isquith IR (1979) In-plant microfouling in desalination. Desalination 30, 337-399

117   Allison DG, Sutherland JW (1987) The role of exopolysaccharides in adhesion of freshwater bacteria. J. Gen. Microbiol., 133, 1319-1327

118   Nivens DE, Chambers JQ, White DC (1990) Monitoring microorganisms at solid surfaces using on-line devices. Abstr. Conf. Microb. Induc. Corr. October 8-12, Knoxville TN

80

119 Hüttinger KJ, Rudi H, Bomar MT (1987) Influence of surface chemistry of the substrate on the adsorption of E. coli. Zbl. Bakt. Hyg. 184, 538-547

120 v. Nägeli C (1893) Über oligodynamische Erscheinungen in lebenden Zellen. Neue Denkschr. allg. Schweiz. Ges. Naturwiss. Abt. 1, 3-51

121 Krause GA (1928) Neue Wege zur Wassersterilisierung (Katadyn). Bermann Publ. Munich

122 Flemming HC (1982) Bacterial growth on ion exchange resins - investigations with a strong cationic exchanger. Part II: Effectiveness of silver against bacterial growth during off-periods of operation. Z. Wasser Abwasser Forsch. 15, 259-266

123 Flemming HC (1984) Bacterial growth on ion exchanger resins - investigations with a strong cationic exchanger. Part III: Disinfection with peracetic acid. Z. Wasser Abwasser Forsch. 17, 229-234

124 Flemming HC, Rentschler H (1983) Testing procedure to proof the sensitivity of water bacteria against silver. Z. Wasser Abwasser Forsch. 16, 157-160

125 Bott TR (1990) Bio-fouling. In: Bohnet M (ed) Fouling of heat exchanger surfaces. Conf. Proc., VDI Ges. P.O.Box 1139, 4000 Düsseldorf 1, Germany; 5.1-5.20

126 Characklis WG (1990) Biofilm processes. In: Characklis WG, Marshall KC (eds) Biofilms. John Wiley, New York; 195-231

127 Turakhia MH, Cooksey KE, Characklis WG (1983) INfluence of a calcium-specific chelant on biofilm removal. Appl. Environ. Microbiol. 46, 1236-1238

128 Sieburth JM, Pratt HL, Johnson PW, Scales D (1978) Microbial seascapes. A pictorial essay on marine microorganisms and their environments. Univ. Park Press, Baltimore; plate 6-76

129 Vrolijk NH, Targett NM, Baier RE, Meyer AE (1990) Surface characteristisation of two Gorgonian coral species: implications for a natural antifouling defence. Biofouling 2, 39-54

130 Wahl M, Banaigs B, Lafargue F (1989) Aufwuchs und Verteidigung oder: Lernen von Meeresorganismen. Spektrum d. Wiss. Feb. 1989, 15-18

BIOFOULING OF REVERSE OSMOSIS MEMBRANES

H.F. Ridgway and J. Safarik

Biotechnology Research Department
Orange County Water District
Fountain Valley, California

ABSTRACT

Reverse osmosis (RO) membranes used for the treatment of industrial and munici-pal process waters often become biologically fouled. The development of a microbial biofilm on the feedwater surfaces of RO membranes results in several adverse effects, including: (i) a gradual decline in the membrane water flux, (ii) an increase in the transmembrane operating pressure [i.e. an increase in the membrane delta-p], and (iii) a reduction in membrane mineral rejection. The RO membrane polymer itself may also be directly or indirectly biodegraded by the adherent microorganisms. Bacterial coloni-zation of the permeate [i.e. product-water] surfaces of RO membranes can also occur. Although the extent of biofilm formation on the permeate surface is typically quite low compared to that on the feedwater surface, it can result in microbial contamination of downstream processes, which may be of great concern in ultra-pure water applications.

Over the last decade, the Orange County water District in southern California has conducted basic and applied research on the mechanism of bacterial adhesion and biofilm formation on RO membranes employed in advanced wastewater treatment. Although this research has been performed principally at Water Factory 21, a 0.66 m³/s wastewater reclamation facility incorporating cellulose acetate [CA] type RO membranes, the general conclusions should extrapolate well to most other RO applications. The primary results of the research are summarized below:

(1) RO Biofilm Bacteria: Early biofilm formation on cellulose acetate membranes used at Water Factory 21 is initiated by acid-fast mycobacteria, which can also be found in significant numbers in the RO feedwater. After some weeks or months of continuous operation, the mycobacteria are eventually replaced by a more diversified microbial community. Other researchers have demonstrated that different types of biofouling bacteria, such as species of *Pseudomonas, Acinetobacter, Staphylococcus* and others may predominate in early biofilm development at RO facilities located elsewhere. The type of biofouling bacteria that predominates at a particular RO facility depends on the physico-chemical and microbiological composition of the feedwater and whether a biocide, such as chlorine, has been added.

(2) Biofilm Growth Rate: Biofilm formation typically occurs in an exponential fashion when a new membrane element is placed into operation. The early increase in microbi-al biomass is correlated with a corresponding decline in RO membrane flux.

H.-C. Flemming · G. G. Geesey (Eds.)
Biofouling and Biocorrosion in Industrial Water Systems
Proceedings of the International Workshop on
Industrial Biofouling and Biocorrosion, Stuttgart, Sept. 13-14, 1990
© Springer-Verlag Berlin Heidelberg 1991

(3) Bacterial Adhesion Kinetics: Laboratory tests indicate that mycobacterial adhesion to RO membranes occurs very rapidly with no discernable lag phase. An initial rapid rate of bacterial adhesion occurring over the first one or two hours is usually followed by a more gradual linear increase in adsorbed cells.

(4) Adhesion Mechanism: Laboratory studies have also shown that the mycobacteria adhere to CA and possibly other RO membrane surfaces primarily by means of a hydrophobic interaction. Consistent with this hypothesis is the observation that adhesion can be largely inhibited by relatively low concentrations of certain non-ionic surfactants. Large changes in the medium pH or other ionic conditions generally result in much smaller inhibitory effects on mycobacterial adhesion. Furthermore, bacteria which exhibit a strongly hydrophobic cell surface, such as the mycobacteria, typically display more rapid adhesion kinetics than hydrophilic bacteria.

(5) Adhesion to Different Membranes: Finally, there appears to be a direct correlation between the extent of mycobacterial attachment and the hydrophobicity of the RO membrane polymer itself. Other properties of the RO membrane which may also influence bacterial adhesion include (i) the magnitude and sign of the membrane charge, (ii) the charge orientation and distribution, (iii) the membrane porosity or density, and (iv) the surface ultrastructure of the membrane.

Several strategies are currently employed to prevent or control microbial biofilm formation in RO systems. These strategies include: (i) refinement of feedwater pre-treatment, e.g. by improving prefiltration or disinfection, (ii) reducing the system operating pressure or recovery, (iii) increasing the frequency of membrane cleaning or improving the cleaning formulation, and (iv) changing the type of RO membrane. Additional research is needed to develop novel RO membrane polymers and module configurations having lower biofouling potentials.

INTRODUCTION

General Background

Reverse osmosis (RO) is a crossflow separation process used most commonly for the demineralization of brackish surface and groundwaters, seawater (11), and municipal wastewater (for wastewater reuse applications; 16). Reverse osmosis also plays essential roles in medical applications (e.g., renal dialysis, 10), ultrapure water production for boiler-feed make up in electric power generation facilities, as well as for the manufacture of pharmaceutical products (14) and semi-conductor devices (9). A wide variety of organic polymers are employed to fabricate RO membranes, including cellulose acetate, aromatic and aliphatic cross-linked polyamide, polyetherurea, polyacrylonitrile, sulfonated polysulfone, and many others (1, 16). Each of these polymers exhibits distinctive physico-chemical properties which determine their respective applications. The membranes can be packaged into various module (i.e. element) configurations, the most common of which are the spiral-wound (Figure 1) and hollow-fine fiber elements (1). All RO membranes function by preferentially excluding dissolved solute molecules while allowing the passage of water (Figure 2); thus, they are considered to be semipermeable (i.e. permselective) membranes (16). Most RO membranes are asymmetric in construction, having one 'active' semipermeable surface in contact with the pressurized feed solution (16). The semipermeable layer is typically very thin (<0.2 micrometers) and is structurally supported by an underlying woven polyester fabric or polysulfone substratum having a high degree of water and solute porosity.

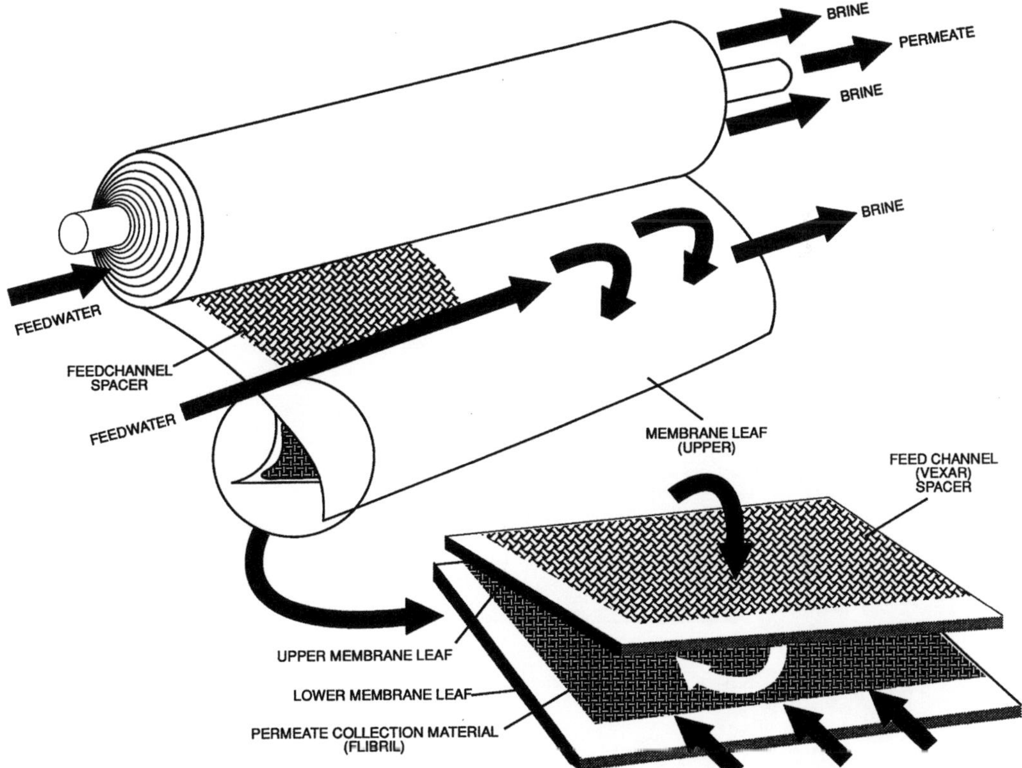

Fig. 1. Schematic of spiral-wound RO module construction. Feedwater flows primarily tangentially to the RO membrane surface.

As alluded to above, the hydrodynamic flow pattern associated with RO membrane elements is primarily tangential to the plane of the membrane surface (Fig. 3). Because individual RO membrane elements are typically operated at relatively low fluid recovery values (usually < 15%) to minimize concentration polarization (16), most of the flow passing through a module is tangential to the feedwater surface of the membrane. Only about 10 - 15% of the total module flow actually passes through the membrane (i.e. normal to the plane of the membrane) and is demineralized. In theory, this type of crossflow pattern should tend to prevent the accumulation of colloids and other suspended particulate matter on the membrane surface. However, in actual practice some colloids and larger particles (including many microorganisms) gain access to the membrane surface where they may become irreversibly immobilized by adsorption processes (17, 18). It is the adsorption of individual bacterial cells and their subsequent growth and proliferation that result in the formation of a microbial biofilm on RO membrane surfaces. Biofilm formation in RO systems is fundamentally similar to surface biofouling phenomena which occur in other natural and industrial environments (5, 28).

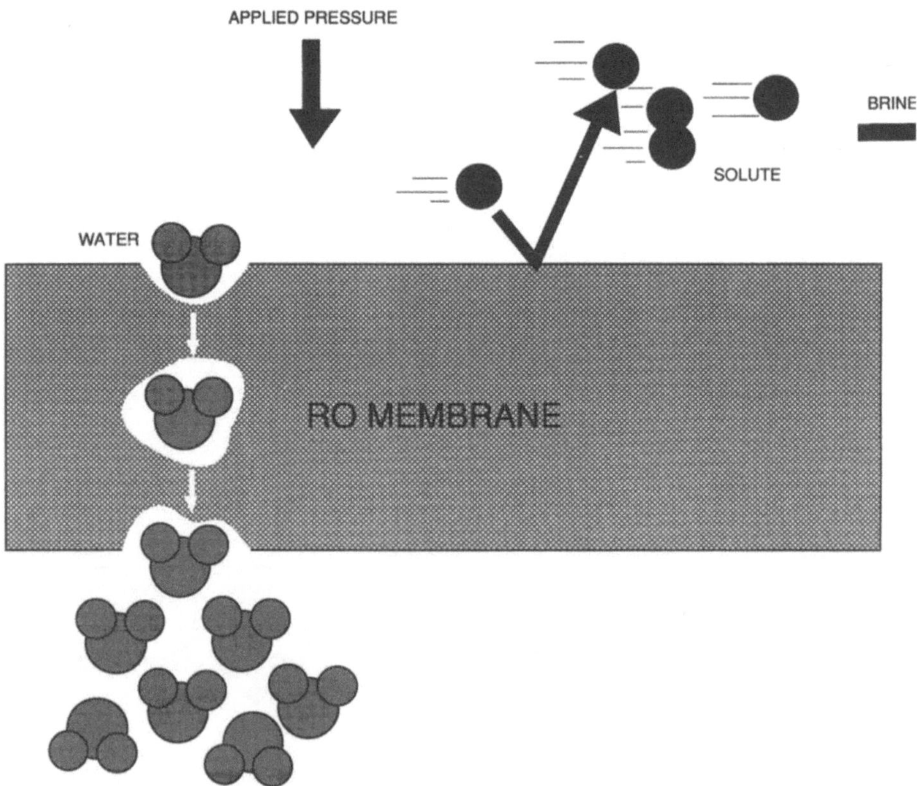

Fig. 2. Schematic illustrating concept of the reverse osmosis diffusion membrane concept

The purpose of this chapter is to briefly review some of the physico-chemical and biological factors influencing microbial adhesion and biofilm formation in RO membrane systems. Additional information concerning this subject may be found in two earlier reviews (17, 18). Some new information will be provided herein regarding the role of the RO membrane surface itself in either promoting or discouraging bacterial attachment.

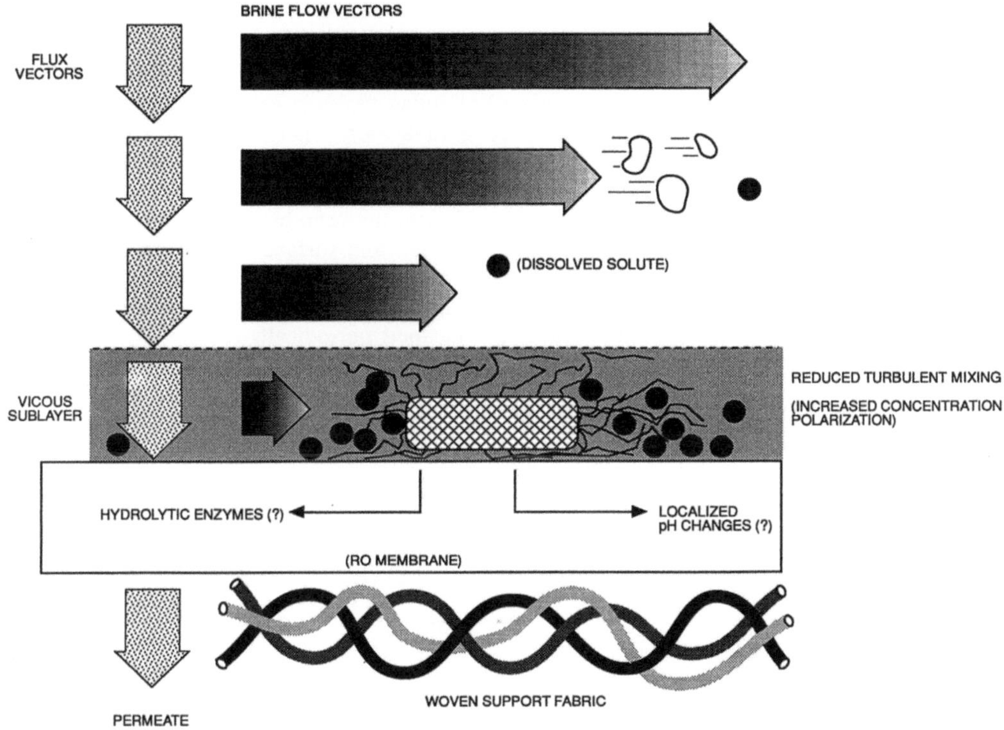

Fig. 3. Schematic illustrating concept of tangential flow in an RO module. The presence of a biofilm and associated extracellular polymers on the feedwater surface of the membrane enhances the opportunity for concentration polarization

Effects of Biofilm Formation

The development of a biofilm on the feedwater surface of an RO membrane results in several undesirable effects which diminish process efficiency and membrane lifetime. These effects, which have been reviewed in more detail elsewhere (17, 18), are briefly summarized below.

Membrane Flux Decline

The accumulation of biological material on the feedwater surface of an RO membrane results in a gradual diminution of water flux through the membrane (17, 18). The decline in water transport activity is not the result of an irreversible change in the composition or structure of the RO polymer, but rather the result of the formation of a low-permeability biological film on the feedwater surface of the RO membrane. It is the biofilm itself (and associated biopolymers) which functions as the primary impediment to molecular water transport through the RO membrane. Partial removal of the biofilm by chemical cleaning typically results in a temporary restoration of water transport, often to near pre-fouling levels. Flux decline due to biofouling typically exhibits biphasic kinetics, an initial rapid decline followed by a more gradual decay (Fig. 4). This characteristic biphasic response evidently results from an initial phase of rapid bacterial adhesion and biopolymer synthesis followed by a period of reduced growth rate. The gradual flux-decay phase may reflect the establishment of an equilibrium between biofilm growth rate on the one hand, and hydraulic fluid shear at the RO membrane surface on the other, although other factors are probably also involved. Because membrane cleaning methods are not completely effective, flux losses due to biofouling often become increasingly irreversible over time, even if the membrane surfaces are cleaned periodically (31).

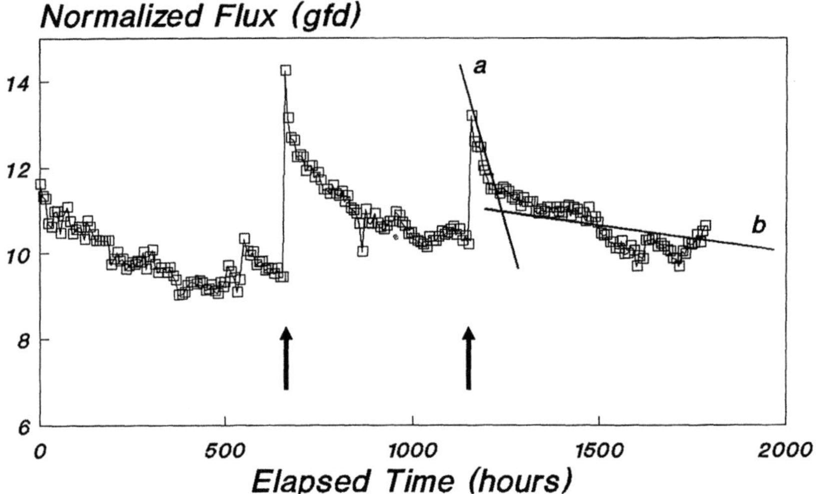

Fig. 4. Example of flux decline data for subunit 2A at Water Factory 21. Note that membrane flux is temporarily restored following each cleaning event (indicated by *arrows.* Also note biphasic flux decline kinetics. *a* = rapid decline phase; *b* = slow decline phase.

Increase in the Transmembrane Operating Pressure

Membrane biofouling frequently results in an increase in the fluid frictional drag in a direction parallel to the long axis of the RO module. This frictional increase may result in substantial energy losses along the length of the RO membrane element which, in turn, cause an increase in pressure dissipation through the element. In severe cases, the transmembrane operating pressure (i.e. the membrane delta-p) may exceed the manufacturer's design specifications for the RO module, resulting in compression and physical collapse along the longitudinal axis of the module (17, 18). Like membrane flux losses which occur due to biofouling, increases in the membrane delta-p can be partially relieved by regular membrane cleaning. It is also sometimes possible to arrest an increase in the membrane delta-p by the introduction of a biocide, such as chlorine, into the feedstream on a continuous or semi-continuous basis (see Fig. 17, below).

Loss of Mineral Rejection

The formation of an attached biofilm is frequently associated with a gradual increase in the solute concentration of the RO permeate water (18). The increase in solute passage through the membrane is generally not the result of an actual change in the permselective properties of the RO membrane. Rather, such losses in mineral rejection are probably an indirect effect produced by an increase in concentration polarization at the membrane surface. The development of a surface biofilm could be expected to extend and stabilize the viscous sublayer (i.e. boundary layer), where dissolved solutes tend to accumulate, thereby leading to an enhanced opportunity for concentration polarization (Fig. 3).

Membrane Biodegradation

Bacteria and other microorganisms comprising the biofilm may, through the excretion of acidic or basic metabolites, produce localized conditions (e.g., pH shifts) which destabilize the RO membrane polymer, making it more susceptible to spontaneous oxidation and/or hydrolytic cleavage. It is also possible that certain bacteria and fungi elaborate specific enzymes which degrade the membrane polymer directly. Several cases of suspected membrane biodegradation have been reported (18, 29).

KINETICS OF BIOFILM FORMATION

Ridgway and coworkers (17-19) demonstrated that bacterial adhesion and biofilm development were surprisingly rapid on the feedwater surfaces of cellulose acetate RO membranes used for demineralizing pretreated, activated-sludge effluent at Water Factory 21 (Figs. 5 and 6). Evidence of biofouling was also observed on the permeate surfaces of RO membranes (20). Typically, an initial rapid increase in the attached bacterial population was followed by a plateau phase where little or no further increase was observed (17-19). The duration of the plateau phase appears to be indefinite, as long as hydrodynamic flows and other system operating parameters remain relatively constant. However, higher horizontal flow velocities, which are associated with lower system recovery values, tend to reduce net biofilm accumulation due to increased tangential fluid shear proximal to the membrane surface.

It is noteworthy that the majority of bacteria comprising an RO biofilm are evidently non-viable, or at least non-culturable. Using direct epifluorescent microscopy to enumerate total (i.e. viable and non-viable) biofilm bacteria, indicates that the number of culturable cells typically ranges from one to several orders of magnitude less than the total cell count (Fig. 5).

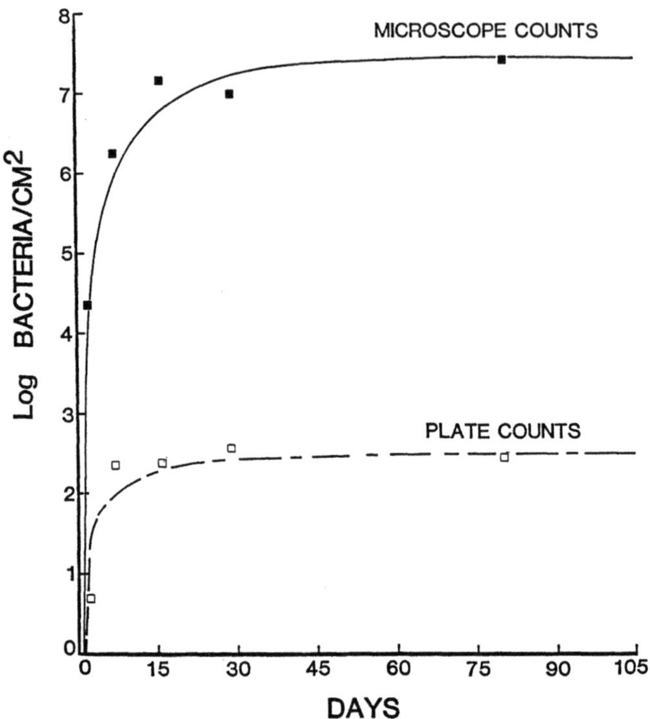

Fig. 5. Kinetics of biofilm development on cellulose acetate RO membranes at Water Factory 21. Note that the microscopic bacterial count is several orders of magnitude higher than the plate counts on R2A medium. The later enumeration method only detects viable (i.e. culturable) cells while the former detects all viable + non-viable cells. Data from Ridgway (17)

The reasons for such a discrepancy (i.e. differential plate count) between viable and total bacterial numbers in the biofilm are not entirely clear. It is possible that many biofilm bacteria are not culturable using commonly employed microbial enumeration media (e.g. R2A medium). However, it is also conceivable that many biofilm bacteria are irreversibly inactivated or otherwise physiologically injured by various biotic (e.g., bacteriophage) or abiotic (e.g., heavy metals) antimicrobial agents accumulated within the biofilm matrix. Presumably, the purposeful addition of a chemical biocide, such as free or combined chlorine, to the feedwater would also tend to increase the differential plate count by establishing highly selective conditions on the RO membrane surface. Under such selective conditions, most bacteria which sorbed to the membrane surface would probably be metabolically inactivated by continuous exposure to the biocide. Only a biocide-resistant subpopulation might be expected to survive and reproduce. It is well documented that chlorine and other microbicidal agents (e.g. heavy metals such as copper) are able to induce physiological injury in bacteria (4, 6, 30), which could, in turn, render a large proportion of the sorbed population non-culturable on standard enumeration media such as R2A.

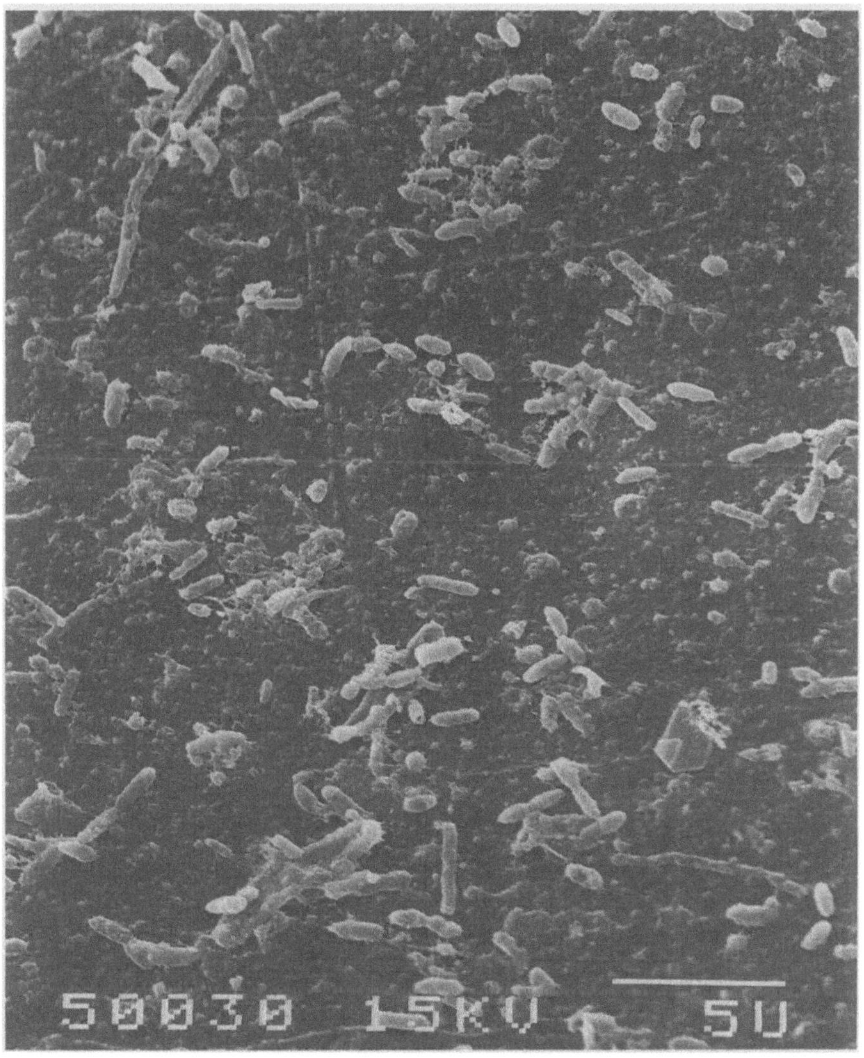

Fig. 6. Biofilm bacteria attached to a cellulose acetate RO membrane operated for a period of six days on a pretreated municipal wastewater feed at Water Factory 21

IDENTIFICATION OF BIOFILM BACTERIA

The identity of bacteria comprising RO biofilms from various locations has been the subject of several investigations in recent years (8, 19, 20, 27). Such information may be useful in designing strategies for biofouling control. Various species of pseudomonades have been most commonly isolated from biofilms growing on RO membranes that have been operated on river and domestic tap water (8, 29). Flemming and Schaule (8) identified *Pseudomonas vesicularis*, *Acinetobacter calcoaceticus*, and *Staphylococcus warneri* as rapidly adhering species from a tap water microflora in southern Germany. Ridgway et al. (20) recovered a total of six different generic groups of bacteria and one unidentified group of microorganisms from cellulose acetate RO membranes fed with pretreated activated-sludge effluent. The six generic groups included *Acinetobacter*, *Pseudomonas/Alcaligenes*, *Bacillus/Lactobacillus*, *Flavobacterium/Moraxella*, *Micrococcus*, and *Serratia*. These bacteria had been isolated from autopsied RO membranes that had been in semicontinuous operation for approximately three years; thus, the biofilm could be considered to have been at a relatively advanced (i.e. mature) stage of development.

Subsequent studies conducted by Ridgway et al. (19) and Safarik et al. (27) involving the collection and analysis of RO membrane scrapings at intervals during biofilm development, revealed that acid-fast mycobacteria were primarily responsible for the earliest colonization of both cellulose acetate and aromatic polyamide membranes used to treat activated-sludge effluent at Water Factory 21 in southern California (Table 1).

Table 1 Occurrence of acid-fast mycobacteria in RO biofilms[*]

Biofilm Age (Days of Operation)	Number of Isolates	Percent Acid-Fast
2	100	100
7	72	100
16	72	98.6
29	73	97.3
80	58	87.9
217	58	79.3
	Total: 433	

[*] All bacteria for acid-fast staining were isolated on R2A medium (Difco). Cellulose acetate membranes operated on pretreated, municipal, activated sludge effluent at the Orange County Water District, Water Factory 21, Fountain Valley, CA

The proportion of acid-fast mycobacteria among feedwater isolates ranged from approximately 13% to 84% during the course of the above investigation (27). The mycobacteria identified were classified as saprophytic 'rapid growers' (e.g. *Mycobacterium fortuitum*) which are not generally recognized as primary human pathogens, although such isolates are sometimes associated with upper-respiratory disease symptoms in immuno-compromised hosts.

The mycobacteria, which were responsible for early biofilm development at Water Factory 21, were eventually replaced months later by a much more diversified microbial community involving species of *Acinetobacter, Shigella, Alcaligenes, Pseudomonas, Klebsiella, Flavobacterium/Moraxella*, as well as gram-positive and unidentified species (19). These observations suggest that biofilm formation undergoes successional changes in microbial community structure.

The environmental signals which result in such changes are undoubtedly complex and have not yet been delineated, even in the broadest terms. It may be speculated that the early-adhering mycobacteria chemically modify the native RO membrane surface through the biosynthesis and deposition of various extracellular biopolymers. Masking of the RO membrane by mycobacterial polymers might render the surface more suitable for subsequent attachment and growth of other biofouling bacteria. Other possible explanations of early biofilm formation by the mycobacteria are: (i) that the mycobacteria are more chlorine-resistant than other biofouling bacteria (a low dose of combined chlorine was, in fact, typically added to the feedwater in the studies described), or (ii) that the mycobacteria may exhibit an enhanced affinity for the RO membrane surface compared to other biofouling microbes (see next section).

THE MECHANISM OF BACTERIAL ADHESION

A strain of mycobacteria (*Mycobacterium BT2-4*), originally recovered from an RO membrane operated for two days at Water Factory 21 (21), has been employed in a series of studies performed by Ridgway and coworkers (21-25) to explore the mechanism of bacterial adhesion to RO membranes (Fig. 7). The bacterial adhesion assay used in these investigations is schematically depicted in Fig. 8 (also see Table 2). Briefly, the mycobacteria are grown for approximately 48 hours at 28°C in a medium supplemented with a mixture of tritiated amino acids to uniformly radiolabel the bacteria. Following radiolabeling, the cells are washed, resuspended in a dilute phosphate buffer solution and placed into contact with the active, semipermeable surface of an RO membrane coupon. After an appropriate contact time, unbound or loosely bound mycobacteria are rinsed off of the membrane coupon and the firmly attached cells are then quantified by determining the amount of membrane-bound radiolabel. The above adhesion assay has proven to be very effective in helping to delineate the kinetics and primary adsorption mechanisms involved in mycobacterial attachment to RO membranes (Fig. 9; 23). By adding various chemical agents to the suspending phosphate buffer solution during an adhesion assay, it is possible to either inhibit or promote bacterial attachment (23).

Inferences can then be drawn concerning the nature of the bonding forces involved in the bacterium-RO membrane polymer interaction. For example, relatively large variations in the pH of the buffer solution resulted in comparatively little inhibition of mycobacterial adhesion to cellulose acetate RO membranes (Fig. 10; 23). Similarly, high concentrations of sodium and lithium salts, as well as anionic polyelectrolytes (e.g., polyethylene imine), did not strongly inhibit adhesion (23). In contrast, a relatively low concentration (0.1% wt/vol) of a non-ionic surfactant (i.e. Triton X-100) tended to strongly interfere with attachment. These and other similar data (23-25) have indicated that mycobacteria probably adhere to cellulose acetate RO membrane surfaces primarily by means of hydrophobic interactions. Polar interactions (e.g. hydrogen and ionic bon-

92

ding) between exposed cell-surface macromolecules and the RO membrane polymer appear to also play a significant but less dominant role in attachment, since large changes in the ionic environment were not completely without effect. The specific biopolymers and other surface ligands involved in mycobacterial adhesion remain unknown. The envelopes of mycobacteria and most other microbes are exceedingly complex (32); consequently, the identification and molecular characterization of adhesive biomolecules exposed at the bacterial cell surface represents a potentially fruitful area of needed research.

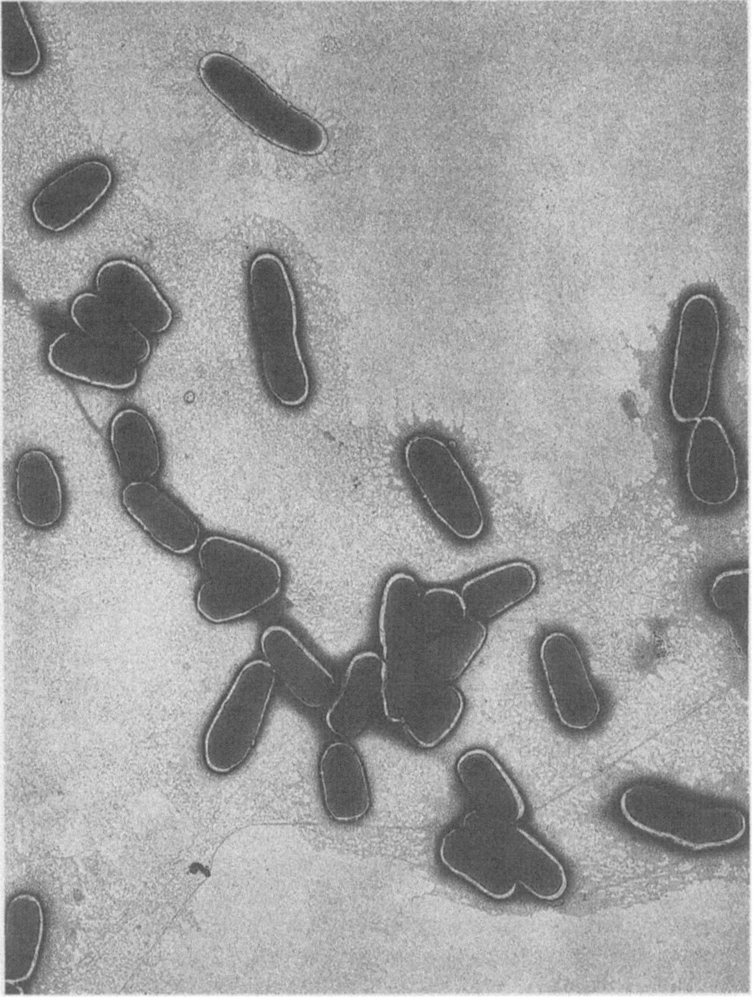

Fig. 7. Transmission electron micrograph of *Mycobacterium strain BT2-4* used in adhesion studies. Note evidence of extracellular polymer associated with the cells

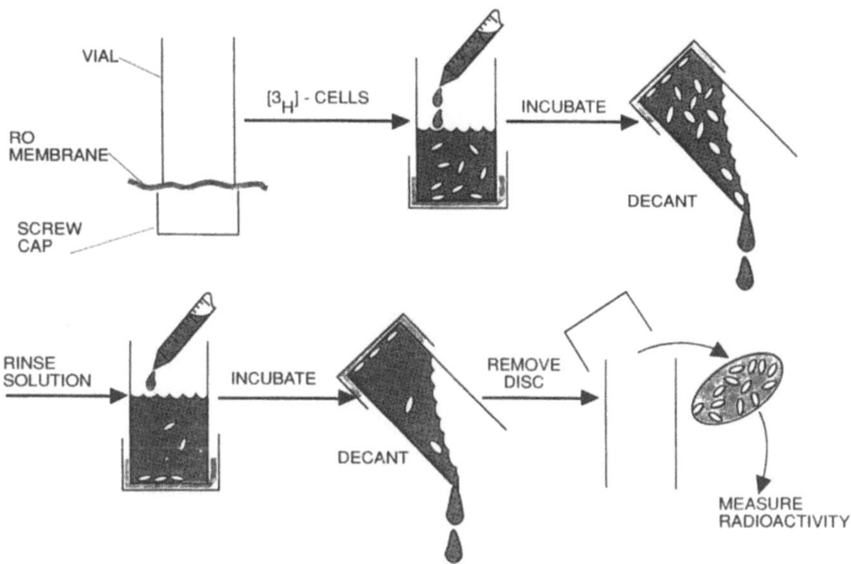

Fig. 8. Schematic of steps involved in the laboratory bacterial adhesion assay. The assay may be used to identify substances that prevent bacterial attachment, or that remove already sorbed cells. Diagram adapted from Ridgway (18)

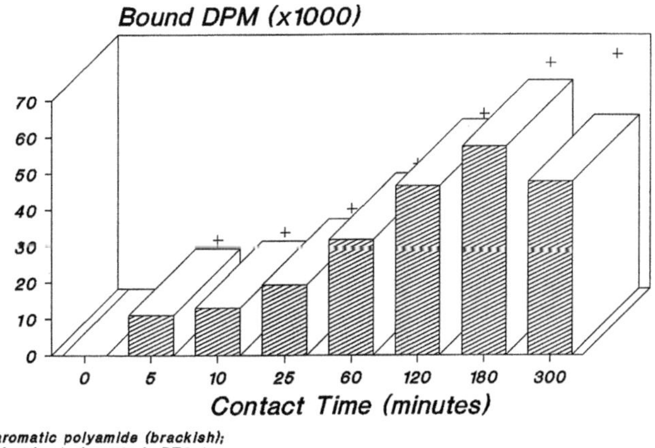

Fig. 9. Kinetics of mycobacterial adhesion to cellulose acetate RO membranes. Bound DPM = membrane-bound disintegrations per minute, a quantitative measurement of attached radiolabeled mycobacteria. Data was obtained using the adhesion assay described in Fig. 8 and Table 2. Data adapted from Ridgway (17)

94

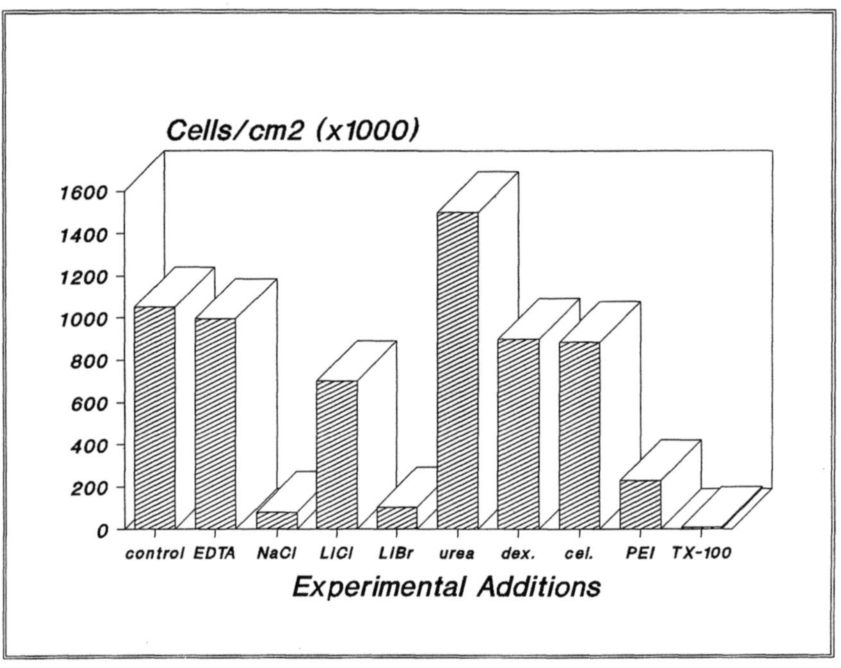

Fig. 10. Effects of various chemical agents on mycobacterial adhesion to cellulose acetate RO membranes. Note strongly inhibitory effect of the non-ionic surfactant Triton X-100. Data adapted from Ridgway et al. (23).

Table 2. Conditions of bacterial adhesion assay used to study microbial attachment to RO membrane surfaces

System Parameter	Description/Value
Buffer System:	10 mM sodium phosphate + 1.0 mM magnesium chloride, pH = 7.0
Assay Temperature:	28°C
Bacterial Concentration:	typically between 5×10^6 and 5×10^7/mL
Assay Volume:	5.0 mL
Assay Duration:	five hours
Other Conditions:	orbital shaking at 200 rpm

Adhesion to Different RO Membranes

The bacterial adhesion assay described above has been more recently employed to explore the interaction of mycobacteria with different RO membrane polymers (8, 18, 25). Large differences have been observed in the rate and extent of mycobacterial attachment to RO membrane polymers having differing chemistries, surface ultrastructures, and porosities (25). Adhesion data presented in Fig. 11 suggest that mycobacterial attachment is directly correlated with membrane hydrophobicity and inversely with membrane hydration capacity, which is consistent with the results of earlier studies described above (23, 24). Little or no significant effect on mycobacterial adhesion was observed when the degree of RO membrane sulfonation was varied extensively (Fig. 12), which is again consonant with previous findings suggesting a minimal influence of polar bonding interactions (23). However, amination of the RO membrane surface resulted in a significant enhancement of mycobacterial adhesion, suggesting that attractive charge interactions can be of importance (Fig. 13; 25). It should be stressed that polar bonding phenomena may be more significant than hydrophobic interactions in the attachment of certain other RO membrane fouling bacteria (e.g. pseudomonads; 8).

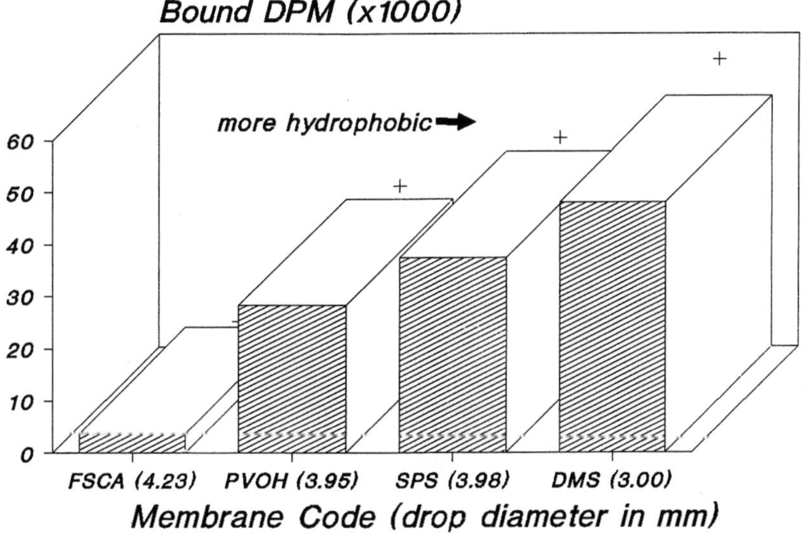

Fig. 11. Influence of RO membrane hydrophobicity on mycobacterial adhesion. Bound DPM (Y-axis) is proportional to the number of membrane-attached radiolabeled mycobacteria. Numbers in parentheses on the x-axis are average diameters of water droplets placed on the dried membrane surfaces. Smaller drop diameters indicate a more hydrophobic membrane. Symbols: *FSCA* = cellulose acetate membrane, *PVOH* = polyvinyl alcohol-coated polysulfone membrane, *SPS* = sulfonated polysulfone membrane, *DMS* = dimethyl silicone

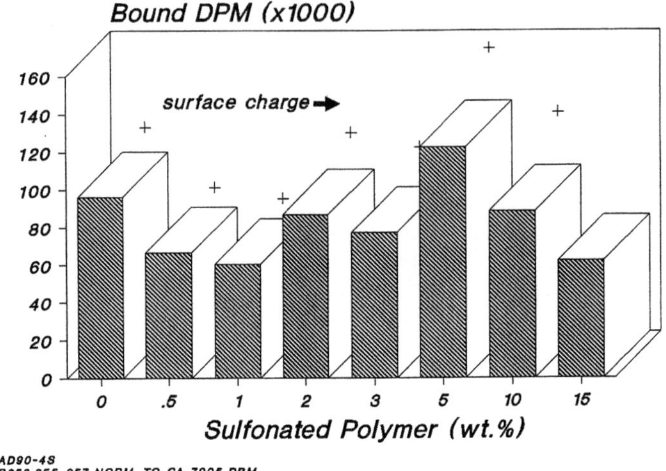

Fig. 12. Influence of RO membrane sulfonation (negative membrane surface charge) on mycobacterial attachment. Y-axis labeling same as in Fig. 11. Note absence of negative surface charge on mycobacterial adhesion

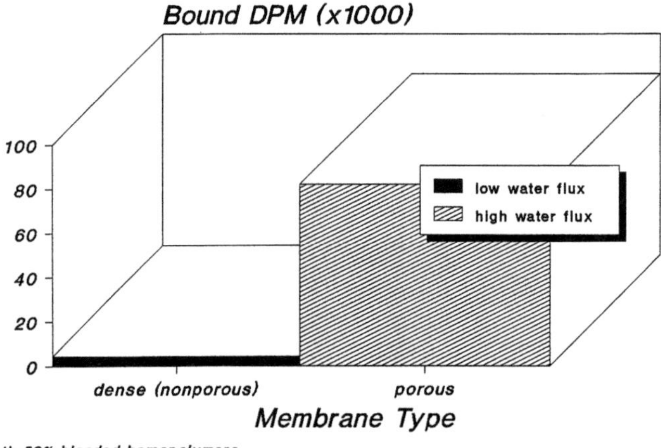

Fig. 13. Comparison of mycobacterial adhesion to two types of sulfonated (i.e. negatively charged) RO membranes and one aminated (i.e. positively charged) membrane type. Numbers in parentheses on X-axis are the percent water content of each polymer type. The higher water content of the aminated membrane reflects greater hydrophilicity which should depress mycobacterial adhesion; however, this factor is offset by the positive charge of the exposed amine groups, which strongly promote attachment. Y-axis labeling same as in Fig. 11

Membrane porosity also appeared to influence mycobacterial adhesion, with more porous RO and ultra-filtration membranes exhibiting a greater degree of mycobacterial attachment (Fig. 14; 30). Unfortunately, the orientation and distribution of the sulfonate groups on this category of membranes varied in an unknown way with the membrane porosity, complicating interpretation of the experimental results. However, assuming the degree of membrane sulfonation does not influence mycobacterial adhesion greatly (as indicated in experiments described above), then it may be inferred that porosity of the RO membrane plays an important role in mycobacterial attachment. It should be noted that the pore structure of all of the RO membranes tested was much smaller than that of a bacterial cell. Thus, increasing adhesion as a function of membrane porosity was probably not associated with physical entrainment of bacteria within membrane pores or crevices. The effects of more subtle surface morphological variations cannot be ruled out at this time and represents an area of continuing research effort. Presumably, a rough membrane surface with shallow depressions and crevices with depths approaching average bacterial dimensions (about one micrometer) would promote microbial adhesion and fouling by establishing localized areas of reduced turbulence and surface shear (Figs. 15 and 16).

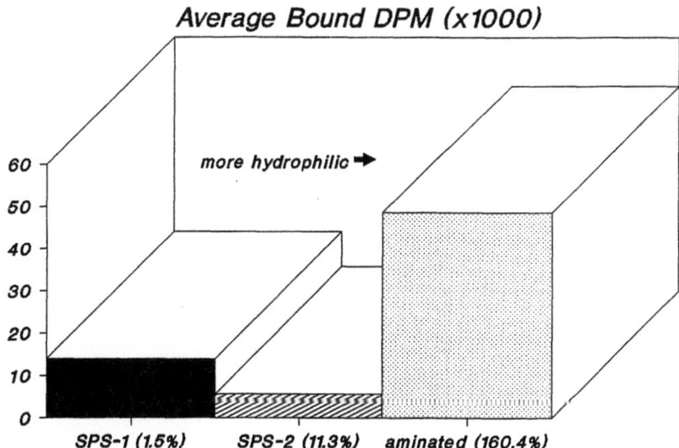

Fig. 14. Influence of membrane polymer density (i.e. porosity) on mycobacterial adhesion. More porous, high water-flux RO polymers increase mycobacterial adhesion for unknown reasons. The membrane pores are not large enough for bacteria to become entrained, ruling out simple physical filtration

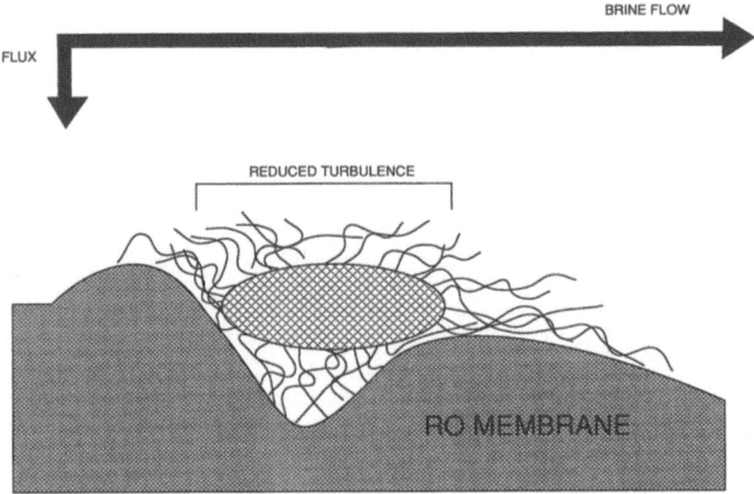

Fig. 15. Diagram illustrating concept of entrainment of a fouling bacterium in a membrane crevice where fluid turbulence could be expected to be low

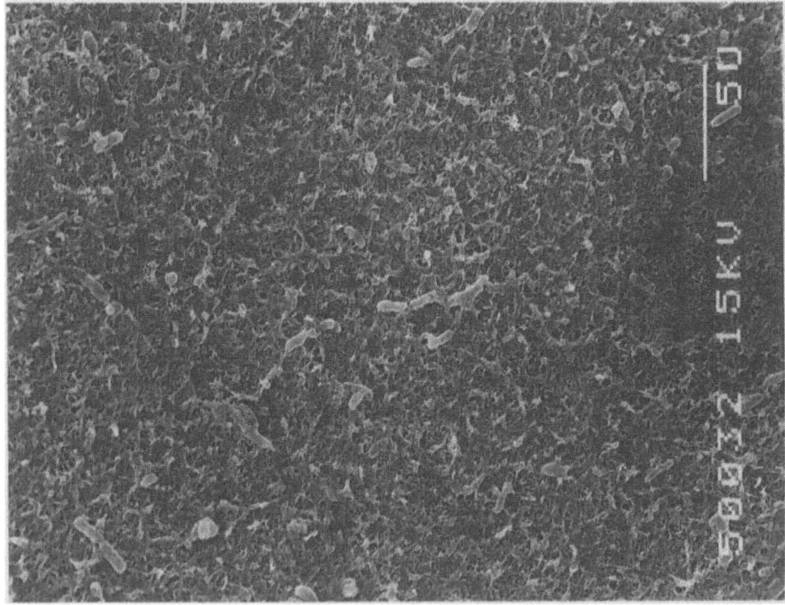

Fig. 16. Scanning electron micrograph showing ruffled surface of an aromatic polyamide RO membrane. Note that bacteria can become sequestered in small irregular crevices in the membrane surface. Bar = 5.0 μm

CONTROL OF MEMBRANE BIOFOULING

There are several fundamental strategies for controlling biological fouling of RO membranes (17). These general strategies, which are listed in Table 3, include: (i) improvement of pretreatment conditions, (ii) changing the type, method, or frequency of membrane cleaning, (iii) modifying system operation, or (iv) changing the type or design of RO membrane.

Table 3. Strategies for the control of RO biofouling

General Category	Specific Strategy
Feedwater Pretreatment:	-improve prefiltration of feedwater -improve preflocculation of feedwater -addition of a feedwater biocide
Membrane Cleaning:	-improve type of cleaning formulation -increase frequency of cleaning -improve membrane cleaning method
System Operation:	-lower system operating pressure -reduce system recovery
Membrane Type:	-change membrane polymer type -change element configuration

Improving RO Pretreatment

In general, it is best to prevent or minimize bacterial attachment and microbial surface growth before it becomes a severe problem. However, in practice this is often not entirely feasible, since only a single viable bacterium entering an RO membrane module could, theoretically, result in extensive biofilm development. Strategies for prevention of RO biofouling through improved pretreatment fall into two broad categories: (i) physical removal of the biofouling bacteria from the RO feedwater, or (ii) metabolic inactivation of the biofouling bacteria by means one or more antimicrobial agents or disinfectants. The currently available technology for continuously removing bacteria from process feed streams remains somewhat limited, primarily because of the tremendous variety and very small size of most bacteria (typically < 1.0 μm) Commonly employed removal strategies include filtration and/or flocculation and settling. The basic problem with these technologies is that filters having pore sizes small enough to remove most bacteria (i.e. <1.0 μm) tend to plug with colloids and microorganisms too rapidly to be economically feasible in large-scale treatment applications. Flocculation polymers sometimes can be used to advantage, but they are frequently unpredictable or inconsistent over time in terms of their performance efficiency. Some of this inconsistency is undoubtedly a reflection of temporal fluctuations which occur in the physico-chemical and microbiological composition of the feedwater.

An untested alternative to conventional filtration or flocculation methods might be the use of high-volume affinity adsorption columns. Theoretically, such columns could be specifically designed for the removal of particularly troublesome biofouling bacteria. For example, it might be possible to design columns packed with a relatively hydrophobic substratum (e.g., dimethylsilicone beads) to preferentially adsorb mycobacteria and other hydrophobic microbes from the RO feedwater. Bacterial removal by affinity adsorption would have the advantage of relatively high initial flow velocities which, presumably, could be restored from time to time by periodic chemical regeneration (e.g. backwashing with detergent solutions). Unfortunately, such columns have not yet been commercially developed and employed for this type of application. Nevertheless, the concept may have practical merit and deserves further investigation and pilot testing.

Perhaps the most effective pretreatment method for impeding membrane biofouling is the continuous or intermittent addition of a chemical biocidal agent to the RO feed stream. It is essential that the biocide selected be economical and sufficiently stable to provide a continuous active residual on the RO membrane surface. Most biocides in use today must also be environmentally compatible and safe for human handling. Although ultraviolet (UV) light disinfection of the feedwater can be quite effective, it does not provide an active biocide residual on the membrane surface, thereby permitting rapid regrowth of any surviving bacteria entering the RO system. On the other hand, a more stable chemical biocide, such as combined chlorine (i.e. monochloramine), not only inactivates microorganisms within the feed stream, but also inhibits cell division and biopolymer synthesis in those bacteria which may have become attached to the membrane surface. By preventing or retarding extracellular biopolymer synthesis, the biofilm may be less stable and more readily removed by membrane cleaning efforts (31).

Selected chemical biocides which have been used to retard biofouling in RO membrane systems are listed in Table 4. Of the compounds listed, combined chlorine (i.e. monochloramine) at concentrations of up to 5.0 mg/L seems to exhibit good overall compatibility with cellulose acetate RO membranes. Surprisingly, combined chlorine has also been successfully used on a continuous injection basis with aromatic polyamide membranes at Water Factory 21 (unpublished observations). LeChevallier and coworkers (12, 13) have recently provided convincing experimental evidence that combined chlorine (i.e. monochloramine) is significantly more effective than free chlorine in terms of its ability to penetrate and inactivate attached biofilms on pipe surfaces. As discussed in a previous section of this chapter, combined chlorine has been used to arrest increases in the delta-p of cellulose acetate RO membranes operated on a municipal wastewater feed (Fig. 17). Presumably, the monochloramine was able to inhibit, but not remove microbial growth within the RO modules, resulting in a cessation in the rise of the delta-p value.

Table 4. Biocidal agents for RO membranes

Biocidal Agent	Typical Concentration	Compatibility with:	
		CA Membrane	TFC Membrane
free chlorine	0.5 - 1.0 mg/L	yes	no
monochloramine	1.0 - 5.0 mg/L	yes	no
chlorine dioxide	0.5 - 2.0 mg/L	yes	no
formaldehyde	5.0 - 25 g/L	yes	yes
glutaraldehyde	5.0 - 25 g/L	yes	yes
isothiazolone	0.1 - 5.0 g/L	yes	yes
bisulfite	10 - 100 mg/L	yes	yes
UV irradiation	> 99.99% kill	1	1
iodine (periodate)	0.1 - 2.0 mg/L	yes	yes
hydrogen peroxide	0.1 - 2.0 g/L	yes	no
ozone	0.5 - 2.0 mg/L	2	2
peracetic acid	0.1 - 2.0 g/L	yes	yes
quaternary amines	0.1 - 5.0 g/L	3	3
sodium benzoate	0.1 - 5.0 g/L	yes	yes
EDTA	0.1 - 5.0 g/L	yes	yes
pH extremes	pH 2 - pH 12	no	yes

[1] UV irradiation may only be used as a feedwater disinfectant; no active disinfectant residual will be left on the RO membrane surface.
[2] Ozone can only be used to disinfect the RO feedwater, since it will damage most RO membrane polymers.
[3] Compatibility of quaternary amines with RO membranes has not been adequately evaluated

102

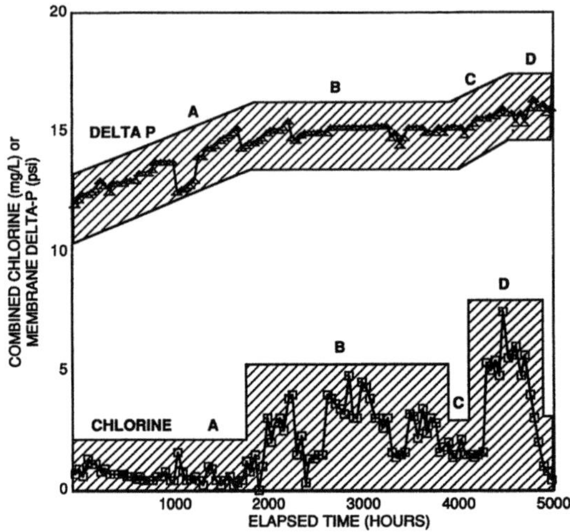

Fig. 17. Inhibition of delta-p increase by introduction of combined chlorine (i.e. monochloramine) into the RO feedwater of Water Factory 21 (subunit 2B data). Note that the delta-p increase was inhibited only when the combined chlorine concentration exceeded approximately 3-4 mg/l (regions *B* and *D* on graphs). When chlorine levels were lower (regions *A* and *C*), the membrane delta-p increased, presumably by resumption of microbial growth in membrane modules

Organic *N*-halamines have shown considerable promise as possible safe replacements for free and combined chlorine in potable water applications (2, 33, 34). Unfortunately, these mildly oxidative compounds have not yet been commercialized and evaluated on a significant scale in RO systems. Isothiazolone ('Kathon'; Rohm and Haas, Inc.) seems to be rapidly replacing formaldehyde and glutaraldehyde as membrane disinfectants for cleaning, storage, and shipping of RO membranes. Both formaldehyde and glutaraldehyde have fallen into disfavor as RO biocides in recent years as concerns over their safety and environmental impacts have grown. However, there are a broad range of related formaldehyde condensate biocides currently available that have not been adequately explored and evaluated for their potential use as RO membrane disinfecting agents and preservatives (26). For reasons which remain somewhat nebulous, the reducing agent sodium bisulfite appears to be a very effective biocide for applications involving seawater demineralization by polyamide membranes. For bisulfite to be effective in seawater RO systems, it should be maintained at a concentration of approximately 50 mg/L, which is close to the minimum concentration needed to establish a completely reducing environment (Robert L. Riley, Separation Systems, International, San Diego, CA, personal communication).

It should be emphasized that application of chemical biocides may do little to prevent biofilm development if the RO membranes are not also cleaned at regular intervals. This is because most biocides alone, including free or combined chlorine, only inactivate the bacteria; they do not necessarily result in cellular lysis and complete destruction and/or physical removal of the target microorganisms (Figs. 18 and 19). Inactivated microorganisms (or their lytic products) can often still adhere to RO membrane surfaces and form a metabolically inactive biofilm (19-24). The accumulated biological material can then serve as a readily available nutrient source for the occasional viable microorganism which escapes biocide inactivation in the feedwater and becomes attached. Such accumulated non-living organic matter can also function in a protective capacity to decrease the effectiveness of biocide activity. Regular membrane cleaning can serve to remove sorbed organic material and minimize the potential for bacterial regrowth and biopolymer synthesis at the membrane surface. Research suggests that the biofilms which develop on RO membrane surfaces that have been continuously exposed to a biocidal agent (e.g. chloramine) are more easily removed by cleaning, presumably because of inhibition of the biosynthesis of extracellular biopolymers involved in the stabilization of the biofilm (31).

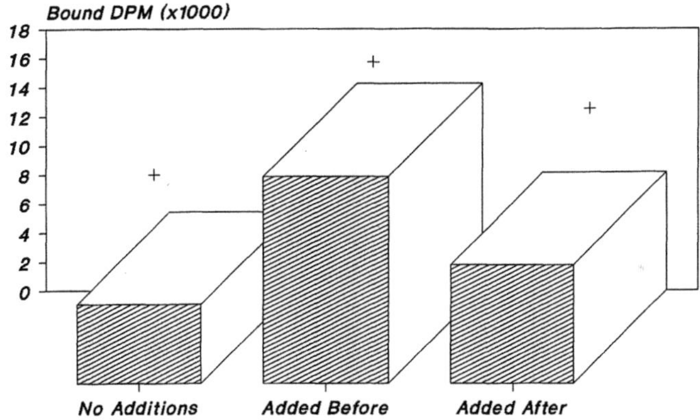

Fig. 18. Influence of free chlorine on mycobacterial adhesion to cellulose acetate RO membranes. A concentration of 1000 mg/L of free chlorine was added either before or after adhesion using the laboratory assay depicted in Fig. 8, above. Note that chlorine addition resulted in greater accumulation of radiolabeled cellular material on the RO membrane surface compared to the control preparation which received no chlorine

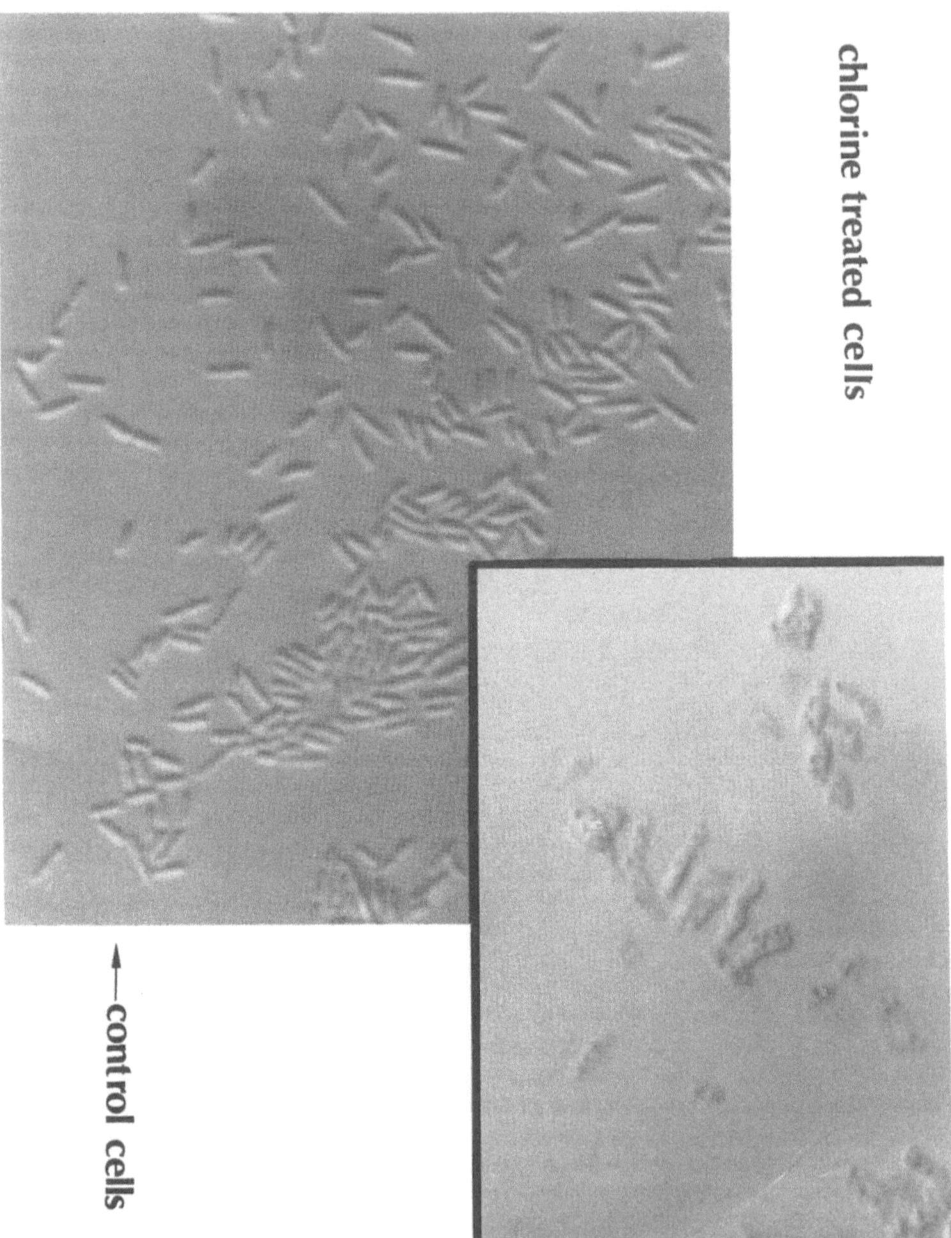

Fig. 19. Nomarski interference contrast light micrograph showing untreated and free-chlorine-treated mycobacteria sorbed onto the RO membranes described in the experiment in Fig. 18. Note lysed appearance of membrane-bound, chlorine-treated mycobacteria

Membrane Cleaning

Cleaning of RO membranes is still a largely empirical science. Because so little fundamental knowledge is currently available concerning the types of cell surface macromolecules and molecular bonding phenomena involved in bacterial adhesion and maintenance of biofilm structural integrity, the precise mechanism of action of most RO membrane cleaning formulations is unknown. The most effective cleaning formulations for biologically-fouled RO membranes generally incorporate one or more neutral or anionically-charged surfactants (8, 17, 31), although certain cationic surfactants, such as the quaternary amines, also show promise as cleaning agents. The later group of surfactants also function as excellent broad-spectrum biocides. Other membrane cleaning components may include: (i) a divalent-cation chelating agent such as ethylenediamine tetraacetic acid (EDTA) or citrate, (ii) organic or inorganic dispersants (e.g. sodium tripolyphosphate), (iii) hydrolytic enzymes (e.g. lipases, proteases, and/or polysaccharidases), and (iv) one or more microbicidal agents (e.g. isothiazolone) which are usually incorporated as a separate treatment step following detergent and enzyme cleaning. A typical RO membrane cleaning formulation is provided in Table 5.

As noted in a foregoing section of this chapter, surfactants are probably involved in the disruption of hydrophobic interactions between the adherent bacteria (or their surface macromolecules) and the RO membrane surface (18). Some of the most commonly used surfactants for RO membrane cleaning are listed in Table 6. It should be cautioned that, for unknown reasons, certain surfactants, such as the quaternary amine dodecyldimethylammonium chloride, may have the potential to enhance bacterial adhesion to RO membrane surfaces (18, 24). Therefore, it is not advisable to assume that any surfactant will be appropriate for incorporation into a cleaning formulation. Each new surfactant must be separately evaluated for its cleaning effectiveness, as well as its possible undesirable side effects on membrane performance.

In order to minimize the adverse effects of biofouling on system performance, the RO membranes must be cleaned at intervals frequent enough to prevent the development of a mature biofilm, which has been shown to be more difficult to remove by cleaning than an early biofilm (31). The exact frequency of membrane cleaning depends on the rate of biofilm growth in a particular system. As a general rule, the RO membranes should be cleaned once per month, or perhaps every other month, when operated under low to moderate biofouling conditions (e.g. demineralization of brackish groundwater or seawater). When operating under more severe fouling conditions (e.g. demineralization of pretreated municipal wastewater), more frequent cleaning will be needed, perhaps once, or even twice per month.

Table 5. Typical formulation for cleaning biofouled RO membranes[*)]

Component	Example	Concentration	Function
surfactant	Triton X-100	0.1% wt/vol	disruption of hydrophobic bonds involved in adhesion; also solubilizes cell membranes and inactivates many bacteria
chelating agent	EDTA	0.01 - 0.1% wt/vol	chelates divalent cations involved in stabilizing cell wall and biofilm integrity
enzymes	polysaccharidase	10 - 100 mg/L	digests extracellular biopolymers involved in cell adhesion and biofilm integrity
dispersant	tripolyphosphate	0.1 - 5.0% wt/vol	assists in solubilizing and suspending particles resulting from disruption of the biofilm by detergents, chelating agents and other factors
microbicide	isothiazolone	0.01 - 0.5% wt/vol	results in metabolic inactivation of biofilm micro-organisms and retards rapid regrowth

[*)] Membrane cleaning done at neutral or alkaline pH (e.g. pH 8.0) at highest allowable temperature and flow. Separate cleaning steps for enzyme and biocide recommended. Each membrane passage should be cleaned separately to minimize cross-contamination by foulants within the system

Table 6. Surfactants commonly used for RO membrane cleaning

Surfactant	Useful Concentration	General Properties
polyoxyethylene ether[*]	0.1 - 10 g/L	non-ionic surfactant; various chain lengths available
sodium dodecylbenzene sulfonate (SDBS)	0.1 - 10 g/L	anionically charged; possibly more effective than Triton series
sodium dodecyl sulfate (SDS)	0.1 - 10 g/L	same as for SDBS

[*] Triton-X and Triton-N series manufactured by Rohm & Haas, Inc.

The various steps involved in an effective membrane cleaning protocol are given in Table 7. Membrane cleaning should be performed using the highest temperatures and flow velocities permissible. In addition, a prolonged soaking phase (minimum of two hours) following the initial membrane rinsing and recirculation phases may be quite beneficial in helping to disrupt and further solubilize the biofilm. A final rinse and soak phase with a chemical biocide (e.g. 0.1% wt./vol. isothiazolone) may also aid in suppressing the rate of biofilm regrowth upon reinitiation of system operation.

Table 7. Steps involved in RO membrane cleaning and sanitization

1. Flush system with permeate for 15 - 30 minutes using maximum allowable delta-p

2. Recirculate surfactant solution at maximum flow allowable for 30 - 45 minutes at 38°C; soak in detergent solution for additional 60 - 120 minutes; chelating agents, such as EDTA, can be added with surfactants

3. Flush system with permeate for 10 - 15 minutes at maximum flow

4. Recirculate enzyme solution (at appropriate pH) for 30 - 60 minutes; soak for 30 minutes

5. Flush system with permeate for 10 - 15 minutes at maximum flow

6. Recirculate with biocide solution for 10 - 30 minutes; soak in biocide for additional 120 minutes

7. Final system flush with permeate for 10 - 30 minutes at maximum flow

System Operation

High operating pressures and system recoveries tend to increase the rate and extent of RO biofouling. Therefore, one strategy to reduce the potential for biofouling is to minimize each of these parameters. The installation of newer high-flux (i.e. low-pressure) RO membranes can be used to not only reduce energy consumption, but also to reduce the degree of membrane biofouling. Greater levels of fluid dynamic shear can be maintained proximal to the RO membrane surface by operating the system at lower recovery values.

Membrane Type or Design

As indicated in a preceding section of this chapter, different RO membrane polymers may exhibit different affinities for specific biofouling bacteria. Presumably, RO membranes which exhibit high bacterial adsorption affinities would also become biofouled more rapidly in actual practice; and experimental evidence seems to corroborate this hypothesis (Fig. 20; 27). If membrane biofouling cannot be adequately controlled by other means, such as biocide addition or improved cleaning, it may become necessary to change the type of RO membrane (e.g. from cellulose acetate to some other polymer material). Spiral-wound cellulose acetate RO modules are, in general, significantly less susceptible to biofouling than hollow-fine fiber membrane configurations; hence, the more frequent selection of the former for most municipal wastewater treatment applications.

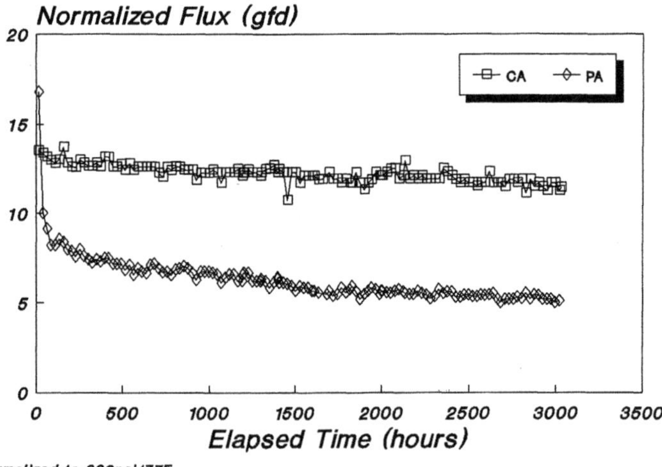

Fig. 20. Normalized flux data for aromatic polyamide (PA) and cellulose acetate (CA) RO membranes operated on a pretreated municipal wastewater feed at Water Factory 21. The polyamide membrane, which evidently fouls more rapidly, also exhibits greater bacterial adhesion in laboratory tests (18, 27)

ACKNOWLEDGEMENTS

The authors wish to thank Mr. Robert L. Riley, Separation Systems International, Inc., La Jolla, California for kindly supplying experimental RO membrane polymers for evaluation in bacterial adhesion experiments. We also thank Ms. Donna Davis, Orange County Water District, Fountain Valley, California for assistance with the artwork.

REFERENCES

1 Allegrezza AE (1988) Commercial reverse osmosis membranes and modules, pp. 53-120. In: Parekh BE (ed) Reverse osmosis technology: applications for high-purity water production. Marcel Dekker, New York, Basel

2 Barnela SB, Worley SD, Williams DE (1987) Syntheses and antibacterial activity of new N-halamine compounds. J. Pharm. Sci. 76, 245-247

3 Block SS (1983) Surface-active agents: amphoteric compounds. In: Block SS (ed) Disinfection, Sterilization, and Preservation. Lea & Febiger, Philadelphia, PA, 335-345

4 Camper AK, McFeters GA (1979): Chlorine injury and the enumeration of water-borne coliform bacteria. Appl. Environ. Microbiol. 37, 633-641

5 Costerton JW, Marrie TJ, Cheng KJ (1985) Phenomena of bacterial adhesion. In: Savage DC, Fletcher MM (eds) Bacterial adhesion: mechanisms and physiological significance. Plenum Press, New York, London, 3-43

6 Domek MJ, LeChevallier MW, Cameron SC, McFeters GA (1984) Evidence for the role of copper in the injury process of coliform bacteria in drinking water. Appl. Environ. Microbiol. 48, 289-293

7 Dychdala GR (1983) Surface-active agents: acid-anionic compounds. In: Block SS (ed) Disinfection, sterilization, and preservation. Lea & Febiger, Philadelphia, PA; 330-334

8 Flemming HC, Schaule G (1988) Biofouling on membranes - a microbiological approach. Desalination 70, 95-119

9 Frith CF (1988) Electronic-grade water production using reverse osmosis technology. In Parekh BS (ed) Reverse osmosis technology: applications for high-purity water production. Marcel Dekker, New York, Basel; 279-310.

10 Ganzi GC (1988) Pure water by reverse osmosis for the treatment of end-stage renal disease. In: Parekh BS (ed) Reverse osmosis technology: applications for high-purity water production. Marcel Dekker, New York, Basel; 399-427

110

11 Ko A, Guy DB (1988) Brackish and seawater desalting. In: Parekh BS (ed) Reverse osmosis technology: applications for high-purity water production. Marcel Dekker, New York, Basel; 185-278

12 LeChevallier MW (1987) Disinfection of Bacterial Biofilms. Proceedings of the Sixth Conference on Water Disinfection: Environmental Impact and Health Effects, May 3-8, pp 1-20.

13 LeChevallier MW, Cawthon CD, Lee RG (1988) Inactivation of biofilm bacteria. Appl. Environ. Microbiol. 54, 2492-2499.

14 Parise PL, Parekh BS, Smith RT (1988) Reverse osmosis for producing pharmaceutical-grade waters. In: Parekh BS (ed) Reverse osmosis technology: applications for high-purity water production. Marcel Dekker, New York, Basel; 347-398

15 Petrocci AN (1983) Suface-active agents: quaternary ammonium compounds. In: Block SS (ed) Disinfection, sterilization, and preservation. Lea & Febiger, Philadelphia, PA.; 309-329

16 Pusch W, Walch A (1982) Synthetic membranes - preparation, structure, and application. Angew. Chem. 21, 660-685

17 Ridgway HF (1987) Microbial fouling of reverse osmosis membranes: genesis and control. In: Mittelman MW, Geesey GG (eds) Biological fouling of industrial water systems: a problem solving approach. Water Micro Associates, San Diego, CA; 138-193

18 Ridgway HF (1988) Microbial Adhesion and Biofouling of Reverse Osmosis Membranes. In: Parekh BS (ed) Reverse osmosis technology: applications for high-purity water production. Marcel Dekker, New York, Basel; 429-481

19 Ridgway HF, Justice CA, Whittaker C, Argo DG, Olson BH (1984) Biofilm fouling of RO membranes-its nature and effect on treatment of water for reuse. J. Amer. Water Works Assoc. 79, 94-102

20 Ridgway HF, Kelly A, Justice C, Olson BH (1983) Microbial fouling of reverse-osmosis membranes used in advanced wastewater treatment technology: chemical, bacteriological, and ultrastructural analyses. Appl. Environ. Microbiol. 45, 1066-1084

21 Ridgway HF, Rigby MG, Argo DG (1984) Adhesion of a *Mycobacterium* sp. to cellulose diacetate membranes used in reverse osmosis. Appl. Environ. Microb. 47, 61-67

22 Ridgway HF, Rigby MG, Argo DG (1984) Biological fouling of reverse osmosis membranes: the mechanism of bacterial adhesion. Proc. Water Reuse Symp. III, The Future of Water Reuse, San Diego, CA, Vol. III, 1313-1350

23 Ridgway HF, Rigby MG Argo DG (1985) Bacterial adhesion and fouling of reverse osmosis membranes. J. Amer. Water Works Assoc. 77, 97-106

24 Ridgway HF, Rogers DM, Argo DG (1986) Effect of surfactants on the adhesion of mycobacteria to reverse osmosis membranes. Proc. Semiconductor Pure Water Conf., San Francisco, CA, 133-164

25 Ridgway HF, Safarik J, Williams J (1990) Microbial interactions with new membrane materials. Proc. Int. Congr. Membranes and Membrane Processes, August 20 - 24, Chicago, Illinois.

26 Rossmoore HW, Sondossi M (1988) Applications and mode of action of formaldehyde condensate biocides. Adv. Appl. Microbiol. 33, 223-277

27 Safarik J, Williams J, Ridgway HF (1989). Analysis of biofilm from reverse osmosis membranes by computer-programmed polyacrylamide gel electrophoresis. Proceedings of the American Society for Microbiology, 14-18 May, New Orleans, LA.

28 Savage DC, Fletcher MM (eds)(1985) Bacterial adhesion: mechanisms and physiological significance. Plenum Press, New York

29 Sinclair NA (1982) Microbial Degradation of Reverse Osmosis Desalting Membranes, Operation and Maintenance of the Yuma Desalting Test Facility, Vol. IV, U. S. Department of the Interior, Bureau of Reclamation, Yuma, Arizona.

30 Singh A, Yeager R, McFeters GA (1986) Assessment of in vivo revival, growth, and pathogenicity of *Escherichia coli* strains after copper- and chlorine-induced injury. Appl. Environ. Microbiol. 52, 832-837

31 Whittaker C, Ridgway HF, Olson BH (1984) Evaluation of cleaning strategies for removal of biofilms from reverse-osmosis membranes. Appl. Environ. Microbiol. 48, 395-403

32 Wicken AJ (1985) Bacterial cell walls and surfaces. In: Savage DC, Fletcher MM (eds) Bacterial adhesion: mechanisms and physiological significance. Plenum Press, New York; 45-70

33 Williams DE, Worley SD, Barnela SB, Swango LJ (1987) Bactericidal activities of selected organic *N*-halamines. Appl. Environ. Microbiol. 53, 2082-2089

34 Worley SD, Williams DE, Barnela SB (1987) The stabilities of new *N*-halamine water disinfectants. Water Res. 21, 983-988

BIOCIDES AND THE CURRENT STATUS OF BIOFOULING CONTROL IN WATER SYSTEMS

Mark W. LeChevallier

American Water Works Service Company, Inc. Belleville Laboratory
1115 S. Illinois St. Belleville, IL 62220, USA

ABSTRACT

Almost all disinfection theory and practice has dealt with inactivation of organisms suspended in a water column. Increasingly, microbiologists recognize that most organisms exist at solid-liquid interfaces. This paper considers disinfection of attached bacteria in potable water systems. Studies detail the mechanisms of disinfection resistance for attached organisms and how these organisms can exist in the presence of a disinfectant residual. Data are presented on the efficiency of various disinfectants for inactivation of biofilm bacteria. The influence of corrosion rates and water chemistries are also examined. A mechanistic model describing the interaction between various biocides and biofilm disinfectant demand shows that faster reacting biocides may be largely consumed before penetrating the biofilm and inactivating attached bacteria. Finally, a combined approach of disinfection and nutrient limitation is presented for optimum biofilm control.

INTRODUCTION

The occurrence of biofilms can cause a multitude of problems in various industrial settings. Biofilms can increase frictional resistance thereby increasing pumping demands and equipment maintenance. Reduced flow rates in occluded pipes can be hazardous in the case of restricted fire protection. Biofilms can decrease heat transfer rates rendering power generating and cooling systems inefficient. Release of bacteria from biofilms can lead to degradation of food, water, beverage and pharmaceutical products. These biofilm organisms can result product-related illness or violation of regulatory standards.

Control of biofilm problems has usually been attempted by application of biocides in the water column. It has been generally assumed that the disinfectant kinetics for biofilm bacteria are similar to those for planktonic organisms. Many industries have experienced biofilm problems, even in the presence of an effective disinfectant residual. For example, it is the experience of the potable water industry that maintenance of a chlorine residual does not always correlate with reduced bacterial counts in the water column (15, 31, 41, 46). It is now recognized that strategies to control of attached bacteria must be based data generated from biofilm experiments. In some cases new theories of biocide-biofilm interaction must be developed. This paper will review strategies for control of biofilm bacteria from the perspective of the drinking water industry. This approach has to consider both the effectiveness of biofilm control and the potability of the water supply. While the number of biocides acceptable for use in potable water are limited, the strategy for biofilm control should be applicable to a variety of industrial settings.

H.-C. Flemming · G. G. Geesey (Eds.)
Biofouling and Biocorrosion in Industrial Water Systems
Proceedings of the International Workshop on
Industrial Biofouling and Biocorrosion, Stuttgart, Sept. 13-14,1990
© Springer-Verlag Berlin Heidelberg 1991

MECHANISMS OF BACTERIAL RESISTANCE

Encapsulation

Several investigators have reported isolation of encapsulated bacteria from chlorinated water (12, 46). These researchers concluded that production of the extracellular capsule helped protect bacteria from chlorine. However, only circumstantial evidence was given to support these conclusions.

Two strains of *Klebsiella pneumoniae* (an encapsulated strain, and a mutant which was incapable of capsule production) were used to examine the impact of capsule production on disinfection efficiency. Growth of the organisms on different media and at various temperatures produced varying amounts of capsular material. Despite these differences in capsular material (as much as a 55 fold), no significant change in the susceptibility of the cells to free chlorine or inorganic monochloramine was observed (32).

Growth of the strains in a low nutrient medium (1 mg/L glucose), however, increased bacterial resistance to free chlorine three fold (Table 1). Evidently, growth under nutrient limiting conditions changed the capsule material. Rudd et al. (49) found that not only did extracellular polymer production increase at low growth rates, but the majority of the capsular material was in a colloidal form of the polymer. At high growth rates, the researchers found that the capsular material was in a more soluble phase. Therefore, the results indicate that possession of an extracellular capsule per se did not increase disinfection resistance, but the form of the capsule (which appeared to be related to growth conditions) affected the bacterial susceptibility to free chlorine. Table 1 shows that neither the growth medium nor capsule production affected disinfection by monochloramine.

Growth Conditions

Growth of the unencapsulated *K. pneumoniae* in low nutrient media increased its resistance to free chlorine two fold (Table 1). Carson et al. (10) reported that Pseudomonas aeruginosa grown in distilled water was markedly more resistant to acetic acid, glutaraldehyde, chlorine dioxide and a quaternary ammonium compound than cells cultured on Tryptic Soy Agar. Similar work by Berg et al. (5) and Harakeh et al. (22) has shown that bacteria grown in a chemostat at low temperatures and submaximal growth rates caused by nutrient limitation, conditions thought to be similar to the natural aquatic environment, were resistant to several disinfectants. *Legionella pneumoniae* grown in a low nutrient "natural" environment have been reported to be 6-9 times more resistant than agar grown cells (28). Berg et al. (5) speculated that the increased resistance was due to changes in the cell membrane permeability of slow growing bacteria. Because the unencapsulated strains shown in Table 1 lacked other defense mechanisms, it is likely that cell membrane changes were responsible for the observed resistance. Olson and Milner (42) presented evidence which showed differences in the cellular protein composition of chlorine sensitive and resistant *Yersinia enterocolitica*.

Table 1. Summary of CxT_{99} values for unattached *K. pneumoniae*

Nutrient level	Encapsulated		Unencapsulated	
	free chlorine	Monochloramine	free chlorine	Monochloramine
High (10 g/l)	0.065 ± .02[a]	32.4 ± 18	0.06 ± .01[b]	31,2 ± 20
Low (1 mg/l)	0.17 ± .05[ab]	29.3 ± 20	0.11 ± .04[b]	37,5 ± 1,0

[a] Statistically different at $p<0.001$
[b] Statistically different at $p<0.05$
Data adapted from (32)

Surfaces

Numerous researchers have shown that increased resistance to disinfection may result from attachment or association of microorganisms to various surfaces including: macroinvertebrates (Crustacea, Nematoda, Platyhelminthes and Insecta) (36, 61), turbidity particles (24, 30, 47), algae (57), carbon fines (9, 34) and glass (41). Ridgway and Olson (47) showed that the majority of viable bacteria in chlorinated water were attached to particles.

Data from Table 1 show that concentration and time (CxT) values to achieve 99 percent inactivation of unattached, high nutrient grown *K. pneumoniae* were 0.065 ± 0.02 mgmin/l for hypochlorous acid and 33 ± 18 mgmin/l for monochloramine. Presented in Fig. 1 are multiples of the CxT units required to inactivate *K. pneumoniae* attached to glass microscope slides. Because of the protection provided by attachment of the bacteria to the glass surfaces, higher multiples (concentration and contact times) were needed to achieve a 99 percent (2 $\log_{10}$) reduction. Data from Fig. 1 show that attachment of *K. pneumoniae* to glass microscope slides increased resistance to hypochlorous acid nearly 150 fold, while resistance of attached organisms to monochloramine increased only two fold.

The results demonstrate that attachment of bacteria to even a relatively inert surface (such as glass) can significantly increase resistance to disinfection. Theoretically, a surface will alter the way a disinfectant interacts with a bacterium by providing steric hindrances to the disinfectant, or by concentration of organic solutes causing a disinfectant demand at the interface. Transport of the disinfectant to the biofilm surface then becomes an important rate limiting step. The importance of mass transfer from the bulk fluid and the diffusion of the compounds within the biofilms has been modeled for several compounds (11, 38). Clearly, understanding the interaction of disinfectants with microorganisms growing on pipeline surfaces is critical to developing effective control strategies.

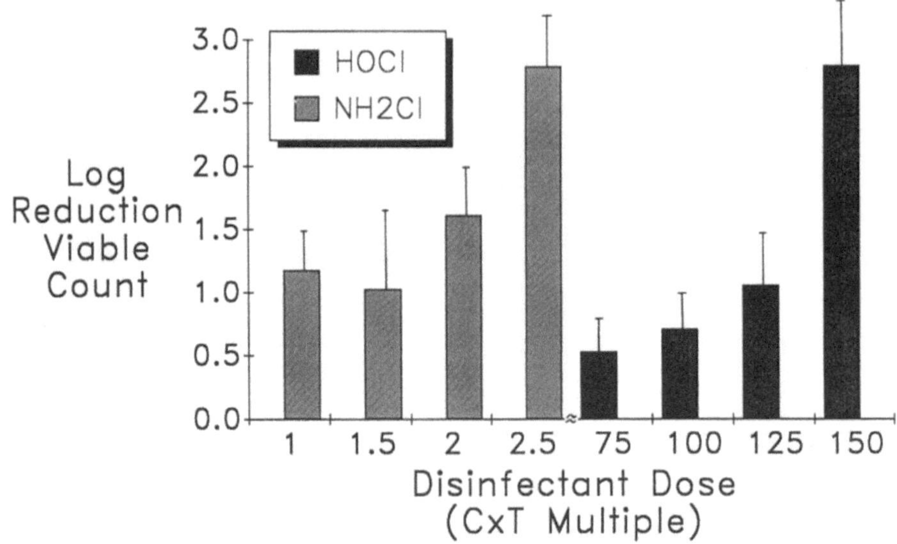

Fig. 1. Disinfection of *K. pneumoniae* attached to glass microscope slides. The log reduction in viable counts is calculated as a comparison with counts for undisinfected controls. Disinfection dose is calculated as multiples of the activity (concentration multiplied by time [CxT], in milligrams per liter) necessary to inactivate 2 $\log_{10}$ (99 %) of unattached bacteria. Used with permission from (32)

Incubation time

Data presented in Fig. 2 indicate that the age of the biofilm may also influence free chlorine disinfection efficiency. The results show that biofilms of the unencapsulated *K. pneumoniae* grown for 7 days in high nutrient media (EPS agar) were 10 times more chlorine resistant than biofilms grown under identical conditions for 2 days. The exact reason for this increased resistance is not known, but it probably was not due to the higher 7 day cell density. Microscopic examination of the biofilms showed that cells were sparsely distributed on the glass surface even after 7 days growth. It is possible that physiological changes in the biofilm population, such as starvation effects, or a coalescence of the biofilm may have made the cells more resistant. Fig. 2 shows that the age of the biofilm, however, did not affect the disinfection efficiency of monochloramine.

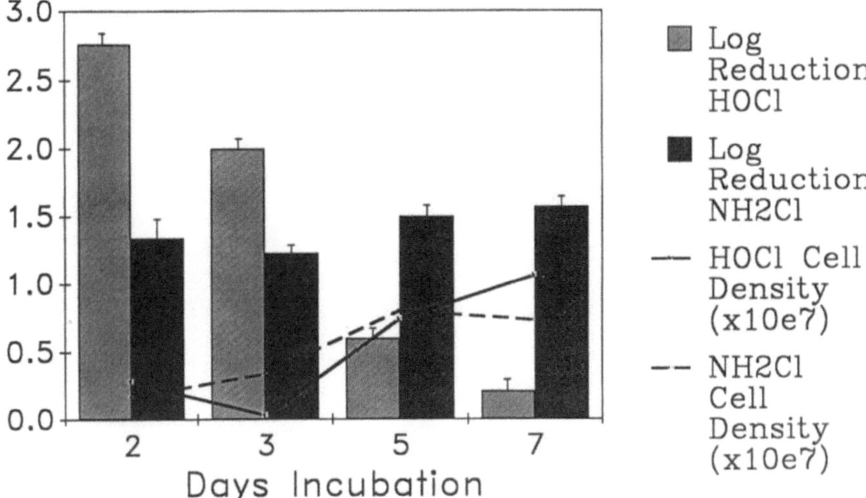

Fig. 2. Effect of incubation time on disinfection. *K. pneumoniae* were grown on glass microscope slides in EPS broth at 35 °C. Biofilms were treated with 1 mg/l free chlorine for 10 min or 5 mg/l monochloramine for 10 min. Bars indicate the $\log_{10}$ (99%) reduction of viable counts to disinfection; lines show the cell density of biofilms on the glass microscope slides. Used with permission from (32)

Cellular Aggregation

Olson and Stewart (43) have reported that aggregation of Acinetobacter strain EB22 increased its resistance to disinfection. The researchers found that aggregation increased resistance of the bacteria to hypochlorous acid over 100 fold, while aggregation increased resistance to monochloramine only 2.3 fold. Sloughing of cell aggregates from treatment filters or pipe walls has been suggested as possible mechanisms by which bacteria can occur in disinfected water supplies.

INTERACTION OF RESISTANCE MECHANISMS

Most studies have examined individual mechanisms of disinfection resistance. It was of interest to know if combined resistance mechanisms were additive or multiplicative. If the resistance of attached *Klebsiella* (150 fold) was additive to the resistance conferred by low nutrient growth (3-4 fold), then the combined resistance would be 153-154 fold. Data presented in Fig. 3 show that the resistance of attached bacteria grown in a low nutrient medium was multiplicative. Bacteria grown under these conditions were approximately 600 fold (150 times 4) more resistant than unattached bacteria grown in rich media.

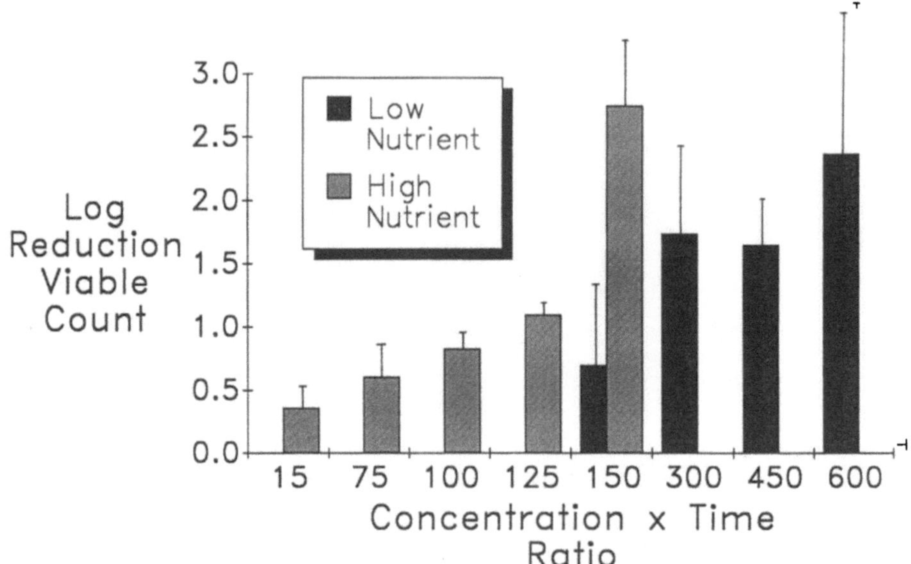

Fig. 3. Multiplicative effect of resistance mechanisms. Data represent combination of resistance provided by growth of *K. pneumoniae* attached to glass microscope slides and resistance provided by growth under low nutrient conditions. Disinfectant dose is calculated as multiples of the activity (in milligram minutes per liter) necessary to inactivate 2 $\log_{10}$ (99%) of unattached bacteria. Used with permission from (32)

This interaction of resistance mechanisms could account for the survival of bacteria in highly chlorinated water supplies. In water distribution systems, encapsulated bacteria attached to pipe surfaces grow under low nutrient conditions for long periods of time, and are usually disinfected using a free chlorine residual. Given this scenario, it is easy to understand how bacteria in biofilms can survive in chlorinated water systems.

DISINFECTANTS FOR CONTROL OF BIOFILM BACTERIA

Data presented in this report show that the choice of disinfectant residual influenced the type of resistance mechanism observed. Disinfection by free chlorine was affected by surfaces, age of the biofilm, encapsulation and nutrient effects. Disinfection by monochloramine, however, was only affected by surfaces.

Disinfection of Biofilms on Metal Coupons

To determine the relative efficacy of various disinfectants for control of biofilm organisms, the CxT coefficient was first determined for unattached bacteria (Table 2). The CxT coefficient represents a unit of activity that would kill 99 percent of the unattached bacteria. For example, 0.08 mgmin/L of hypochlorous acid (pH 7.0; 1-2 °C) would kill 99 percent of the suspended bacterial population (Table 2). Likewise, under the same conditions, 94 mgmin/L of monochloramine would have the same activity. Application of equal CxT units would have one of three possibilities:

*   Bacteria attached to surfaces would be disinfected with equal effectiveness as dispersed organisms (i.e. one CxT unit would inactivate 99 percent of the attached bacteria)
*   The surface would impair the effectiveness of all disinfectants to the same extent (e.g. to some degree less than 99 percent)
*   Given the equal activities of the various disinfectants, one biocide will inactivate attached bacteria better than the others.

Comparison of equal activities (equal CxT units) of hypochlorous acid, chlorine dioxide or monochloramine on the same bacteria grown on metal coupons showed that monochloramine was more effective for inactivation of biofilm bacteria (Fig. 4). One CxT unit reduced viable counts 1.4 $\log_{10}$, while equivalent activities of hypochlorous acid or chlorine dioxide had little effect. To match the disinfection activity of 1 CxT unit of monochloramine, it required 100 CxT units of chlorine dioxide and over 600 CxT units of hypochlorous acid.

Based on the CxT data for unattached cells, one would predict that monochloramine would be 1,200 fold less effective that hypochlorous acid (Table 2). However, experiments performed on biofilm bacteria showed a totally different picture.

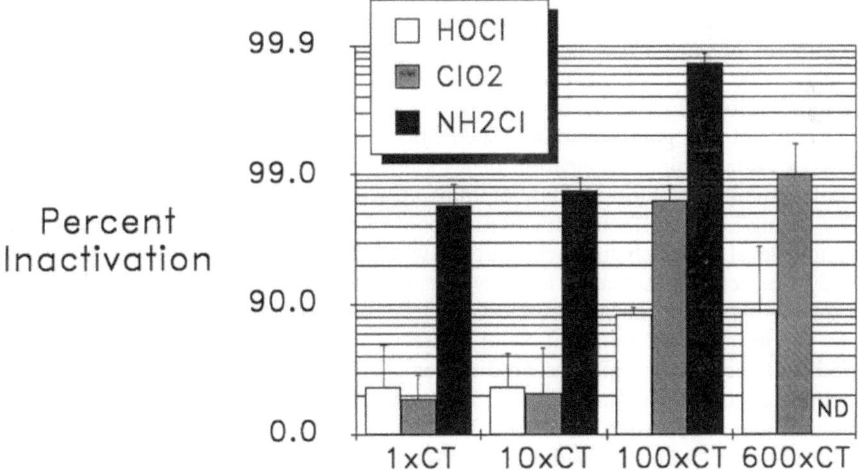

Fig. 4. Inactivation of biofilms grown on metal coupons. Disinfectants were compared on the basis of equal activity (equal CxT coefficients). The solution was stirred using a magnetic stir bar (pH 7,0 at 4 °C). ND=not determined. Used with permission from (32)

Table 2. Comparative efficiacy of disinfectants for the production of 99 percent inactivation in demand-free systems

Disinfectant agent	pH	C x T
Hypochlorous acid	7.0	0.08 ± 0.02
Hypochlorite	8.5	3.3 ± 1.0
Chlorine dioxide	7.0	0.13 ± 0.02
	8.5	0.19 ± 0.06
Monochloramine	7.0	94.0 ± 7.0
	8.5	278.0 ± 46

All CxT calculations (milligrams per liter) are based on a minimum of two trials. Each trial was performed in triplicate at 1-2°C. Data adapted from (33)

Disinfection of Biofilms in a Model Pipe System

Comparisons of free and combined chlorine for control of biofilm bacteria were performed in a model pipe systems under chemical and hydraulic conditions similar to a water distribution system. The model system was 72 ft (29.95 m) long and composed of 1 ft (30.5 cm) sections of 0.5 in (1.27 cm) diameter iron, galvanized, copper, and PVC pipe joined together with quick-fit couplings to provide easy access to the interior of the pipe system for analysis of the biofilm populations (Fig. 5).

Treatment of galvanized, copper or PVC pipe surfaces with either free chlorine or monochloramine at 1 mg/L or 4 mg/L doses (pH 7.0, 25 °C) reduced biofilm counts more than a hundredfold (Fig. 6). For copper and PVC surfaces, application of a free chlorine residual resulted in a greater biofilm inactivation than a similar dosage of monochloramine. For galvanized pipe, application of a monochloramine residual resulted in a greater inactivation than free chlorine. For biofilms grown on iron pipe surfaces, only monochloramine (4 mg/L) was effective for biofilm control. Additional studies showed there was a threshold level at which monochloramine was effective for control of biofilms on iron pipe (35). In actual practice, it is likely that the threshold level will vary, depending on water quality and pipe characteristics. The mechanism of action of monochloramine may account for its effective penetration of bacterial biofilms. Jacangelo and Olivieri (25) reported that monochloramine reacted rather specifically with nucleic acids, tryptophan, and sulfur- containing amino acids. The researchers observed no reaction between chloramines and sugars such as ribose. LeChevallier et al. (32) showed that monochloramine did not react with extracellular polysaccharides. Free chlorine, however, is known to react with a wide variety of compounds (8, 40, 60). This difference in specificity may allow monochloramine to penetrate the biofilm surface and react with the microorganisms, whereas free chlorine is largely consumed before it can fully penetrate the biofilm surface. Selection of biocides, specific for microbial processes and not consumed by side reactions, would be effective choices for biofilm control.

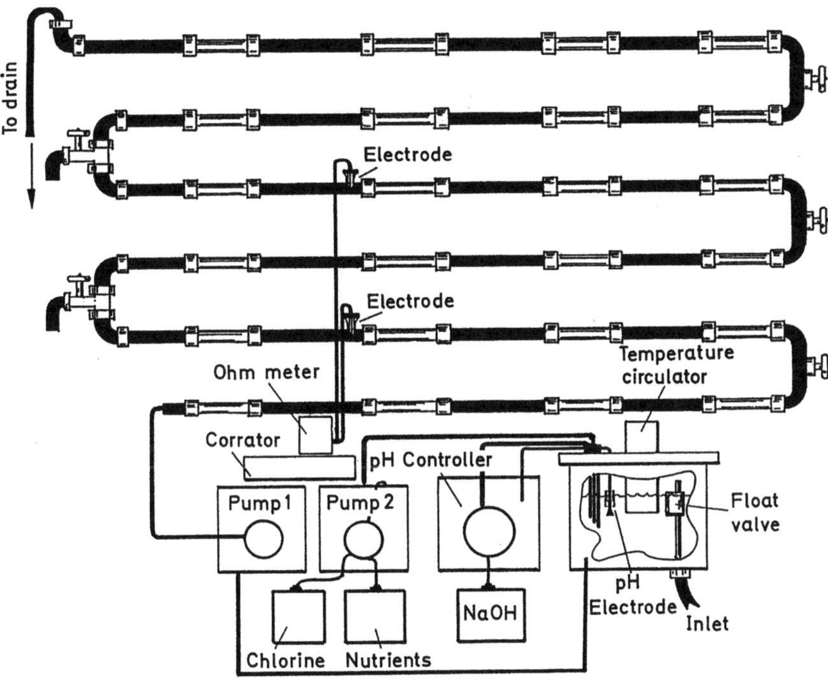

Fig. 5. Diagram of the model pipe system. Used with permission from (35)

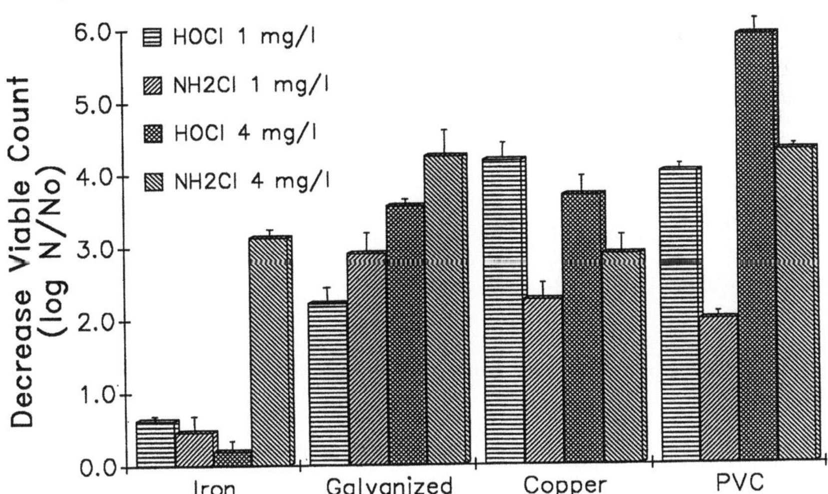

Fig. 6. Summary of biofilm disinfection experiments. Biofilms were grown for two weeks then exposed to the disinfectant for an additional two week interval. The decrease in viable counts after disinfection were calculated as the difference from the initial viable count. Used with permission from (35)

IMPACT OF CORROSION AND WATER CHEMISTRY

The observation that iron pipe surfaces protect attached bacteria from inactivation by a free chlorine residual suggests that corrosion interferes with biofilm disinfection efficiency. Chlorine is known to react with ferrous iron to produce the insoluble ferric hydroxide (66) as follows:

$$2Fe(HCO_3)_2 + Cl_2 + Ca(HCO_3)_2 ---> 2Fe(OH)_3 + CaCl_2 + 6CO_2$$

Each part of iron as Fe oxidized requires 0.64 mg/L of chlorine. White (66) indicated that if iron is present in a complex form, free chlorine is more effective than combined chlorine in breaking up the iron complex so that ferrous oxidation can proceed. It is likely that biofilm organisms complex ferrous ions from metal surfaces within the glycocalyx layer. Free chlorine, therefore, not only reacts with the extracellular polysaccharides, but liberates ferrous ions which also consume the chlorine residual. It is interesting to note that in several systems (31, 44) high levels of biofilm bacteria were associated with iron tubercles.

Application of corrosion inhibitors can improved the disinfection efficiency of free chlorine for biofilm bacteria on iron pipes (35). Results summarized in Table 3 show that application of polyphosphate, zinc orthophosphate, and pH and alkalinity adjustment resulted in improved (10 - 100 fold) biofilm disinfection by free chlorine.

Table 3. Impact of corrosion on disinfection of biofilm bacteria. Data adapted from (35)

Treatment	Corrosion rate (mm/y)	Biofilm inactivation
Carbonate	14-19	< 0.5
None	12-14	< 0.5
Phosphate buffer	ND	1.2
pH adjustment	1-3	1.0
Zinc orthophosphate	1	1.5
Polyphosphate	1	2.0

[a] Values are expressed as $\log_{10}$ reduction of viable counts for tree chlorination (3-4 mg/l residual). ND=not determined

Current research is examining the effect of varying water chemistry on disinfection of biofilm bacteria. Preliminary results show that compounds including moderate levels (20 to 50 mg/L) of sulfate and chloride can decrease disinfection efficiency. These compounds are known to increase corrosion rates of iron, copper and galvanized steel (52). In addition, high sulfate levels can enhance densities of sulfate reducing bacteria. These anaerobic bacteria can stimulate corrosion and create a reduce environment which is extremely difficult to treated with chemical oxidants. In addition, accumulation of cations including sodium, potassium and lithium have been shown to alter the chemical and disinfectant properties of hypochlorite (20).

MODELING DISINFECTANT - BIOFILM INTERACTIONS

Haas et al. (21) have modeled the interaction between free and combined chlorine and bacterial biofilms. The Chick-Watson equation:

$$\ln(N/N_o) = -k'C^n t$$

was found to adequately describe the inactivation kinetics of suspended bacteria. Incorporating the Chick-Watson equation into a model (Table 4) which described the liquid mass transport resistance of a disinfectant to the biofilm surface and the loss of the disinfectant due to demand reactions occurring within the biofilm, four parameters were determined:

x_1 = the rate of reaction/rate of liquid transport $(k\sigma S_o/k_1)$
x_2 = the second order rate of reaction of sites $(k\sigma)$
x_3 = the Chick-Watson rate constant (k')
x_4 = the Chick-Watson coefficient of dilution (n)

Table 4. Model of chlorine-biofilm interaction. Data adapted from (21)

Model of steady state $(dC/dt = 0)$ mass transfer of chlorine into the biofilm:

$$(C/S_b) = 1/[1 + (k\sigma S_o/k_1)(S/S_o)] \qquad (1)$$

Model of mass balance of reaction sites within the biofilm:

$$d(S/S_o)/dt = -k\alpha(S/S_o)C \qquad (2)$$

where: S_o and S are the initial and time-dependent concentrations of reactant sites
k_1 = liquid mass transfer coefficient in the biofilm (length/time)
σ = biofilm thickness (length/time)
k = second order reaction constant (volume/mass-time)
C_b = liquid concentration of disinfectant
α = reaction site:chlorine stoichiometric ratio

Equation (2) is substituted into Eq. (1) and simultaneously solved with the Chick-Watson equation (3) to calculate the surviving fraction (N') in the biofilm:

$$d \ln(N')/dt = -k'C^n \qquad (3)$$

Because x_3 and x_4 were determined from the suspended growth experiments, x_1 and x_2 could be computed. Data shown in Table 5 indicate that for both HPC bacteria and *Klebsiella*, free chlorine had a substantially greater ratio of biofilm disinfectant demand to mass transfer (x_1) than monochloramine. For *Klebsiella*, the reaction rate parameter (x_2) was less for monochloramine than for free chlorine.

The model identifies the disinfectant-demand kinetics of the biofilm to be the key component which dictates the survival of attached bacteria. Accumulation of disinfectant-demand substances (e.g. corrosion products, reduced compounds, microbial products, etc.) can create an environment suitable for microbial growth despite large concentrations of biocides in the water column. Selection of biocides which have few disinfectant-demand reactions, even if the biocides are relatively weak, can ultimately achieve inactivation of biofilm bacteria. Alternatively, eliminating the accumulation of disinfectant-demand substances can improve the effectiveness of highly reactive biocides.

Table 5. Summary of pseudo-steady-state parameters for biofilm disinfection model. Data adapted from (21)

Experiment[a]	x^1	x^2	x^3	x^4
Klebsiella				
Free chlorine	7.65×10^8	2.54×10^{-7}	0.0472	0.99
Monochloramine	1.864	4.53×10^{-8}	3.43×10^{-3}	0.90
HPC Bacteria				
Free chlorine	1.41×10^3	1.31×10^{-6}	0.59	0.94
Monochloramine	0.97	1.86×10^{-3}	6.52×10^{-4}	1,01

[a] For explanation of terms, see text

NUTRIENT CONTROL

Oxidants added for disinfection purposes may increase the amount of biodegradable organic material in treated water. In this manner, biocides added to kill biofilms actually help increase the amount of nutrients available for growth. Ozonation of waters containing complex, long- chain, biorefractory components causes the conversion of these compounds into more readily biodegradable substances (2, 4, 26, 37). Hargesheimer et al. (23) reported that ozonation increased the concentration of both aliphatic and halogenated acids in treated water supplies. Van der Kooij (62) reported that chlorine disinfection significantly increased the biodegradable organic carbon content of water.

Data presented in Fig. 7 show a 2.2 fold increase in assimilable organic carbon (AOC) levels after raw water is ozonated. The increase in AOC is primarily due to stimulation of Spirillum strain NOX; a bacterium capable of utilizing oxylated carbon compounds (62, 63). After treatment through GAC filters, AOC levels were reduced 85 percent. However, post-chlorination increased AOC levels 66% In this instance, chlorination stimulated both Pseudomonas fluorescens strain P17 and strain NOX.

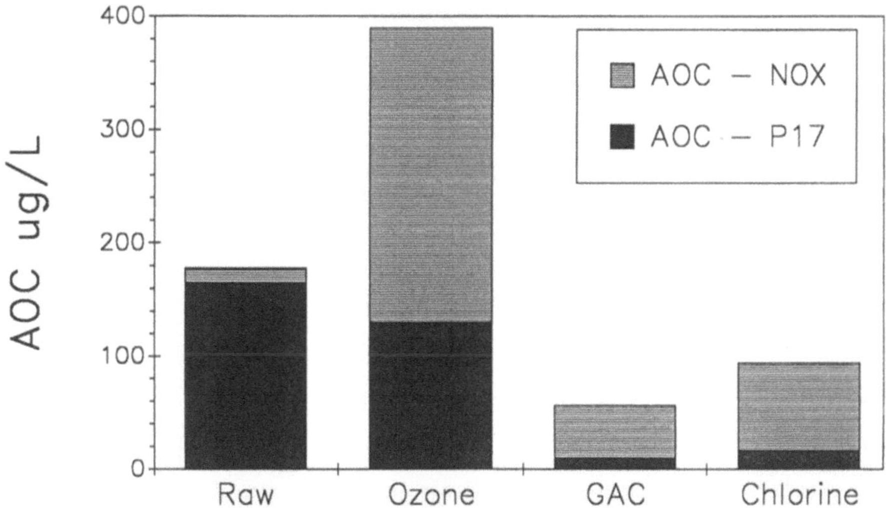

Fig. 7. Impact of disinfection on assimilable organic carbon (AOC) levels in water. Raw water was ozonated with a 2.5 mg/l dose for 15 min contact time. After granular activated carbon (GAC) filtration, the water was dosed with 2.0 mg/l free chlorine for 30 min. AOC levels were determined according to the method of Van der Kooij (63)

In some cases, chemicals added to treat process water (eg. coagulants, corrosion inhibitors, ammonia etc.) can add nutrients for bacterial growth. Because most heterotrophic bacteria require carbon, nitrogen and phosphorous in a ratio of approximately 100:10:1, carbon is often the growth-limiting compound. However, in circumstances of nitrogen or phosphorous limitation, addition of ammonia (for chloramination) or phosphate (as a corrosion inhibitor) could have serious consequences. It is advisable, therefore, to characterize the water supplies to determine the growth-limiting factors.

Removing the nutrients which are required for growth of attached bacteria can limit the size and activity of biofilms. Biological treatment is the practice by which microbial activity is encouraged within the treatment process sufficient to remove all nutrients which might support significant bacterial growth in treated effluents. Application of biological processes may involve many forms. Biological removal of organic material have been reported for aerated submerged media reactors (56), fluidized bed filters (18, 27, 55), slow sand filters (16, 51), rapid sand filters (6, 16, 58, 65) and granular activated carbon filters (3, 13, 39, 65). Various studies have shown that biological treatment can be highly effective, with more than 90% removal of biodegradable organic material (3, 6, 14, 17, 26, 45, 48, 59, 63). Bourbigot et al. (6) showed that installation of ozone treatment followed by biologically active carbon filtration effectively reduced nutrients and controlled biofilm problems in the Mery-sur-Oise distribution system.

Van der Kooij (62-64) has shown that assimilable organic carbon levels of <<10

126

µg/L can limit bacterial growth in treated waters. In fact, Van der Kooij (65) and Schellart (50) report that biological stability can be so effective, that some utilities in the Netherlands have eliminated secondary disinfection as a means of controlling bacterial growth in distribution system water.

Biological treatment can also improve the stability of certain biocides. Bablon et al. (2) showed that biologically active carbon filtration combined with ozonation synergistically reduced the chlorine demand of the treated water. The lower chlorine demand produced a more stable chlorine residual in the distribution system. Bablon et al. (3) reported that rechlorination stations within the distribution network were not needed due to the improved chlorine stability of biologically treated water. The lower chlorine demand of biologically treated water may reduce chlorine by-product formation. A study of GAC operations at the Jefferson Parish, LA, water treatment plant showed that microbial activity was responsible for steady state removal (50 percent) of organic precursors to chloroform production (7). These results may be important to industries which discharge treated process water.

DEVELOPMENT OF A STRATEGY FOR BIOFILM CONTROL

A strategy for biofilm control should include consideration of (i) the engineered system; (ii) the water chemistry, including nutrient concentration; (iii) the microorganisms to be treated, and; (iv) the biocide. For effective biofilm control each of these areas must be addressed. It is not realistic to expect that biocide application alone will be sufficient to control biofilms in a poorly designed and maintained system.

The Engineered System

The engineered system should have smooth surfaces, free of pits or other obstructions which might provide protection for attached bacteria. The system should be designed with good circulation and, if possible, turbulent flow. Stagnation, even on a microscopic level, will lead to bacterial growth and water quality degradation. The engineered system should be composed of materials which do not contribute to a disinfectant demand or provide nutrients for bacteria growth. Solvents in some paints and resins, and rubber and synthetic materials used in joints, gaskets, and coatings have been shown to promote biofilm formation (1, 19, 53, 54).

Water Chemistry

Only recently has attention focused on the impact of water chemistry on biofilm formation and control. Corrosive water can lead to accumulation of products on the pipe surface which can create a disinfectant demand and provide a habitat for bacterial growth. Methods are just being developed to quantitate biodegradable compounds in water. Current research is developing a standardized and simplified method for measuring assimilable organic carbon in water. Clearly, more information is needed to understand the relationship between water chemistry and biofilm development.

Microorganisms

Microorganisms have at their disposal a variety of resistance mechanisms to avoid or interfere with disinfection. Through the use of confocal scanning laser microscopy scientists are now beginning to study structure of biofilms (29). Anaerobic layers within a biofilm can produce a reduced environment which is most difficult to disinfect. Metabolic activities within the biofilm can reduce compounds including iron, sulfate and nitrate by use as electron acceptors. These compounds can act as catalysts with disinfectants through a cycle of chemical oxidation and microbial reduction. Reducing biofilm activity through nutrient limitation can help break the disinfectant-demand cycle. Removing the compounds by chelation or physical disruption may also improve disinfection.

Biocides

More research is necessary to describe the interaction between biocides and biofilm interfaces. Because disinfectant-demand compounds can be concentrated in biofilm layers, research must consider the role of biocide consuming side reactions. More important than the biocidal efficiency, is the transport and concentration of a disinfectant in the biofilm itself. Given time, even a weak biocide can kill attached organisms. Development of better disinfectant models could allow the selection of the appropriate biocide for a particular biofilm problem.

ACKNOWLEDGMENTS

The author thanks Ramon G. Lee and Richard H. Moser for their comments and suggestions. This research was supported by the funds from the American Water System.

REFERENCES

1   Ashworth J, Colbourne JS (1986) Microbial alterations of drinking water by building services materials - field observations and the United Kingdom water fittings testing scheme. Proc Biodeterioration Society Meeting, Deft Holland

2   Bablon G, Ventresque C, Damez F, Hascoet MC (1986) Removal of organic matter by means of combined ozonation/BAC filtration, a reality on an industrial scale at the Choisy-Le-Roi treatment plant. Proc AWWA Annual Conf, June 22-26, Denver, CO

3   Bablon G, Ventresque C, Roy F (1987) Evolution of organics in a potable water treatment system. Aqua 2, 110-113

4   Baozhen W, Jun Y, Jinzhi T, Qixiang F, Renfen L, Junli H (1985) A preliminary study of the efficiency and mechanism of THM removal in the ozonation and BAC process. Ozone Sci. Eng. 6, 261- 273

128

5   Berg JD, Matin A, Roberts PV (1982) Effect of antecedent growth conditions on sensitivity of Escherichia coli to chlorine dioxide. Appl. Environ. Microbiol. 44, 814-819

6   Bourbigot MM, Dodin A, L'Herritier R (1982) Limiting bacterial aftergrowth in distribution systems by removing biodegradable organics. Proc. AWWA Annual Conf., Miami Beach, FL

7   Brodtmann NV, DeMarco J, Greenberg D (1980) Critical study of large-scale granular activated carbon filter units for removal of organic substances from drinking water. In: McGuire MJ, Suffet, IH (eds), Activated Carbon Adsorption of Organics From the Aqueous Phase, Ann Arbor Science Publishers, Ann Arbor, MI

8   Burleson JL, Peyton GR, Glaze WH (1980) Gas- Chromatographic/Mass-Spectro-metric analysis of derivatized amino acids in municipal wastewater products. Environ. Sci. Technol. 14, 1354-1359

9   Camper AK, LeChevallier MW, Broadway SC, McFeters GA (1986) Bacteria associated with granular activated carbon particles in drinking water. Appl. Environ. Microbiol. 52, 434-438

10   Carson LA, Favero MS, Bond WW, Peterson NJ (1972) Factors affecting comparative resistance of naturally occurring and subcultured *Pseudomonas aeruginosa* to disinfectants. Appl. Environ. Microbiol. 23, 863-869

11   Chang HT, Rittmann BE (1987) Mathematical modeling of biofilm on activated carbon. Environ. Sci. Technol. 21, 273- 280

12   Clark TF (1984). Chlorine tolerant bacteria in a water distribution system. Public Works 115, 65-67

13   Committee Report, AWWA (1981) An assessment of microbial activity on GAC. J. Amer. Water Works Assoc. 73, 447-454

14   Crowe PB, Bouwer EJ (1987) Assessment of Biological Processes in Drinking Water. AWWA Research Foundation Denver, CO

15   Earnhardt KB Jr (1980) Chlorine resistant coliform-the Muncie, Indiana experience. Proc. AWWA Water Quality Tech. Conf., Miami Beach, FL

16   Eberhardt M, Madsen S, Sontheimer H (1977) Investigations of the use of biologi-cally effective activated,carbon filters in the processing of drinking water. EPA--TR-77-503, US Environmental Protection Agency, Cincinnati, OH

17   Faust SD, Aly, OM (1987) Adsorption Processes for Water Treatment. Butter-worths Publ., Boston, MA

18   Foster, JD (1972) Biological treatment of R Trent water. Water Treatment Exam. 21(4), 327- 333

19   Frensch K, Hahn JU, Levsen K, Nieben J, Scholer HF, Schoenen D (1987) Solvents from the coating of a storage tank as a reason of colony increase in drinking water. Vom Wasser 68, 101-109

20   Haas CH, Keralius MG, Brncich DM, Zapkin MA (1986) Alteration of chemical and disinfectant properties of hypochlorite by sodium, potassium, and lithium. Environ. Sci. Technol. 20, 822-826

21   Haas CN, LeChevallier MW, Geoffry M (1990) Modeling of chlorine inactivation of drinking water biofilms. Water Res., submitted

22   Harakeh MS, Berg JD, Hoff JC, Matin A (1985) Susceptibility of chemostat-grown *Yersinia enterocolitica* and *Klebsiella pneumoniae* to chlorine dioxide. Appl. Environ. Microbiol. 49, 69-72

23   Hargesheimer EE, Irvine GA, Badakhshan A, Seidner RT (1986) Pilot studies into effects of disinfection strategies on drinking water quality. In: Huck PM, Toft P (eds) Treatment of Drinking Water for Organic Contaminants, Pergamon Press, New York; 135-149

24   Hoff JC (1978) The relationship of turbidity to disinfection of potable water. In: Hendricks CW (ed) Evaluation of the Microbiology Standards for Drinking Water, EPA-570/9-78-OOC, US Environmental Protection Agency, Washington, DC; 103-117

25   Jacangelo JG, Olivieri VP (1987) Mechanisms of inactivation of microorganisms by combined chlorine. AWWA Research Foundation, Denver, CO

26   Janssens JG, Meheus J, Dirickx J (1984) Ozone enhanced biological activated carbon filtration and its effect on organic matter removal, and in particular on AOC reduction. Water Sci. Tech. 17, 1055-1068

27   Jeris JS, Owens RW, Hickey R, Flood F (1977) Biological fluidized treatment for BOD and nitrogen removal. J. Water Pollut. Control Fed. 49(5), 816-831

28   Kutcha JM, States SJ, McGlaughlin JE, Overmeyer JH, Wadowsky RM, McNamara AM, Wolford RS, Yee RB (1984) Enhanced chlorine resistance of tap water adapted *Legionella pneumophilia* as compared with agar medium passed strains. Appl. Environ. Microbiol. 50, 21-26

130

29   Lawrence JR, Korber DR, Hoyle B, Costerton JW, Caldwell DE (1990) Visualization of microbial biofilm architecture using confocal scanning laser microscopy. Q-28, p. 293 Abstracts Ann. Meet. Amer. Soc. Microbiol.

30   LeChevallier, MW, Evans TM, Seidler RJ (1981) Effect of turbidity on chlorination efficiency and bacterial persistence in drinking water. Appl. Environ. Microbiol. 42, 159-167

31   LeChevallier MW, Babcock TM, Lee, RG (1987) Examination and characterization of distribution system biofilms. Appl. Environ. Microbiol. 53, 2714-2724

32   LeChevallier MW, Cawthon CD, Lee RG (1988) Factors promoting survival of bacteria in chlorinated water supplies. Appl. Environ. Microbiol. 54, 649-654

33   LeChevallier MW, Cawthon CD, Lee RG (1988) Inactivation of bacterial biofilms. Appl. Environ. Microbiol. 54, 2492-2499

34   LeChevallier MW, Hassenauer TS, Camper AK, McFeters GA (1984) Disinfection of bacteria attached to granular activated carbon. Appl. Environ. Microbiol. 48, 918-923

35   LeChevallier MW, Lowry CD, Lee RG (1990) Disinfecting biofilms in a model distribution system. J. Amer. Water. Works Assoc. 82(7), 87-99

36   Levy RV, Cheetham RD, Davis J, Winer G, Hart FL (1984) Novel method for studying the public health significance of macroinvertebrates occurring in potable water. Appl. Environ. Microbiol. 47, 889-894

37   Maloney SW, Suffet IH, Bancroft K, Neukrug HM (1985) Ozone- GAC following conventional US drinking water treatment. J. Amer. Water Works Assoc. 77:8, 66-73

38   Matson JV, Characklis WG (1976) Diffusion into microbial aggregates. Water Res. 10, 877- 885

39   Miller GW, Rice RG (1978) European water treatment practices - the promise of biological activated carbon. Civil Eng. 48, 2-81

40   National Research Council (1980) Drinking Water and Health. Volume 2, National Academy Press, Washington DC

41   Oliveri VP, Bakalian AE, Bossung KW, Lowther ED (1985) Recurrent coliforms in water distribution systems in the presence of free residual chlorine. In: Jolley RL, Bull RJ, Davis WP, Katz S, Roberts MH Jr, Jacobs VA (Eds) Water Chemistry, Environmental Impact and Health Effects. Lewis Publishers, Inc, Chelsea, MI; 651-666

42   Olson, BH, Milner BB (1990) Development of disinfectant- resistant organisms. In: LeChevallier MW, Olson BH, McFeters GA (eds) Assessing and Controlling Bacterial Regrowth in Distribution Systems Amer Water Works Assoc, Denver, CO; 49-117

43   Olson BH, Stewart MS (1990) Factors that change bacterial resistance to disinfection. In: Jolley RL, Condie LW, Johnson JD, Katz S, Minear RA, Mattice JS, Jacobs VA (Eds) Water Chemistry, Environmental Impact and Health Effects. Lewis Publishers, Inc, Chelsea, MI; 885-904

44   Opheim D, Grochowski J, Smith, D (1988) Isolation of coliforms from water main tubercles. N-6 Abst., Ann. Meet. Amer. Soc. Microbiol. p. 245

45   Pascal O, Joret JC, Levi Y, Dupin T (1986) Bacterial aftergrowth in drinking water networks measuring biodegradable organic carbon (BDOC). Proc Ministere de l'Environnement/US Environmental Protection Agency Franco-American Seminar, Oct 13-17, Cincinnati, OH

46   Reilly JK, Kippen JS (1983) Relationship of bacterial counts with turbidity and free chlorine in two distribution systems. J. Am. Water Works Assoc. 75 309-312

47   Ridgway HF, Olson BH (1982) Chlorine resistance patterns of bacteria from two drinking water distribution systems. Appl. Environ. Microbiol. 44, 972-987

48   Rittmann BE, Huck PM (1989) Biological treatment of public water supplies. Crit. Rev. Environ. Control 19(2), 119-184

49   Rudd T, Sterritt RM, Lester JM (1982) The use of extraction methods for the quantification of extracellular polymer production by *Klebsiella aerogenes* under varying cultural conditions. European J. Appl. Microbiol. Biotechnol. 16, 23-27

50   Schellart JA (1986) Disinfection and bacterial regrowth: some experiences of the Amsterdam water works before and after stopping the safety chlorination. Wat. Supply 4, 217-225

51   Schmidt K (1979) Experience with the removal of micro- impurities in slow sand filters. In: Kuhn W, Sontheimer H (eds), Oxidation Techniques in Drinking Water Treatment. EPA-570/9-79- 020, US Environmental Protection Agency, Cincinnati, OH; 620-646

52   Schock MR (1990) Internal corrosion and deposition control. In: Pontius FW (ed) Water Quality and Treatment. McGraw-Hill, Inc, New York; 997-1111

53   Schoenen D, Scholer HF (1985) Drinking Water Materials: Field Observations and Methods of Investigation. Ellis Horwood Ltd, Chichester, England

54   Schoenen D, Wehse A (1988) Microbial colonization of water by the materials of pipes and hoses: changes in colony counts. Zbl. Bakt. Hyg. B 186, 108-117

55   Short CS (1975) Removal of ammonia from river water -2, Technical Report TR3, Water Research Center, Medmenham, England

56   Sibony J (1982) Development of aerated biological filters for the treatment of waste and potable water. Proc 14th congress AIDE, Sept 6-10, Zurich, Switzerland

57   Silverman GS, Nagy LA, Olson BH (1983) Variations in particulate matter, algae, and bacteria in an uncovered, finished drinking water reservoir. J. Am. Water Works Assoc. 75, 191-195

58   Sontheimer H, Heilker E, Jekel MR, Nolte H, Vollmer FH (1978) The Mulheim process. J. Amer. Water Works Assoc. 70:7, 393-396

59   Sontheimer H, Hubele C (1988) The use of ozone and granular activated carbon in drinking water treatment. In: Huck PM, Toft P (eds), Treatment of Drinking Water for Organic Contaminants. Pergamon Press, New York; 45-66

60   Symons JM, Stevens AA, Clark RM, Geldreich EE, Love OT Jr, DeMarco J (1981) Treatment techniques for controlling trihalomethanes in drinking water. EPA-600/2-81-156, US Environmental Protection Agency, Cincinnati, OH

61   Tracy HW, Camarena VM, Wing F (1966) Coliform persistence in highly chlorinated waters. J. Am. Water Works Assoc. 58, 1151-1159

62   Van der Kooij D, Hijnen WAM (1985) Measuring the concentration of easily assimilable organic carbon (AOC) treatment as a tool for limiting regrowth of bacteria in distribution systems. Proc AWWA Water Tech Conf, Houston, TX

63   Van der Kooij D, Visser A, Hijnen WAM (1982) Determining the concentration of easily assimilable organic carbon in drinking water. J. Amer. Water Works Assoc. 74:10, 540-545

64   Van der Kooij D, Visser A, Oranje JP (1982) Multiplication of fluorescent pseudomonads at low substrate concentrations in tap water. Antonie van Leeuwenhoek 48, 229-243

65   Van der Kooij D (1987) The effect of treatment on assimilable organic carbon in drinking water. In: PM Huck and P Toft (eds), Proc Second National Conference on Drinking Water. Edmonton, Canada, April 7-8, 1986, Pergamon Press, London; 317-328

66   White GC (1986) Handbook of Chlorination. 2nd ed, Van Nostrand Reinhold Co, New York; 368-369

BACTERIAL GROWTH AND BIOFOULING CONTROL IN PURIFIED WATER SYSTEMS

Marc W. Mittelman

Institute for Applied Microbiology at the
University of Tennessee, 10515 Research Drive
Suite 300 Knoxville, TN 37932, USA

ABSTRACT

Purified water used in product formulation, cleaning, and cooling operations has evolved from a process fluid to an essential raw material. Levels of biological and abiological contaminants which were undetectable five years ago are now regarded as unacceptable for many products and processes. While recent advances in analytical chemistry and fine particle physics have resulted in dramatic reductions in abiological contaminants, biological contamination of purified waters used in these critical industries remains a significant challenge to future product development. In the semiconductor industry, 1 μm and smaller line-width devices are subject to fatal defects as a result of bacteria present in "ultrapure water." Active ingredient degradation and pyrogen contamination of heat labile biological and medical devices have been linked to purified water contamination. The extent of bacterial growth and biofilm formation in 18 MOhmcm waters is a function of materials selection, systems design, and preventative maintenance protocol. Limiting essential growth factors - C, N, P, S, trace elements, light - in purified water systems is an important key to the control of biological contamination. Development of on-line, real time biofouling detection systems is currently underway. These evolving systems, which include supercritical fluid extraction of signature biomarkers and electrochemical impedance spectroscopy should provide insight into conditions promoting the development of fouling biofilms. Future applications of novel detection and treatment systems will include advanced life support systems such as those found on the space station.

INTRODUCTION

The use of purified water in various industrial and medical applications has increased dramatically over the past twenty years. These different applications often require varying levels of water quality. Each industry sets specifications for the acceptance of purified water quality based upon their product or process demands. The aggressive nature of ionically ultrapure water (18.2 Mohm·cm) can contraindicate its use in, for example, stainless steel distribution systems which are often employed in the pharmaceutical industry. While the presence of trace levels of silica in condensate polishing loop waters creates great concern in the power industry, low-level silica contamination of purified waters used in the production of medical devices or photomasks, for example, may not constitute cause for alarm. Bacterial pyrogens, or

H.-C. Flemming · G. G. Geesey (Eds.)
Biofouling and Biocorrosion in Industrial Water Systems
Proceedings of the International Workshop on
Industrial Biofouling and Biocorrosion, Stuttgart, Sept. 13-14,1990
© Springer-Verlag Berlin Heidelberg 1991

134

endotoxins, are of serious concern to ethical pharmaceutical operations and medical device manufacturers; however, their significance as contaminants in power generation and semiconductor purified water systems is unresolved. If any one group of contaminants can be viewed as "universal" in their distribution, significance, and recalcitrance, they are the bacteria and their by-products. Their role as purified water contaminants appears to cross all boundaries of purified water application and usage.

In the semiconductor industry, the demand for contaminant free water has, to a great extent, driven ultrapure water technologies. Indeed, water purification technologies have advanced to the point whereby levels of ionic, organic, and particulate contaminants can be reduced to concentrations below current analytical detection limits. Unfortunately, the industry's ability to detect and remove biological particulates and organics has not kept pace with the needs of today's VLSI devices. Yang and Tolliver (1) have noted that, at the 1-megabyte level where minimum circuit feature size is typically 1 μm, device yields are limited by 0.1 μm and larger particles.

Bacteria may represent the greatest fraction of total particles in this size range. Current analytical capabilities for ultrapure water analysis have generally limited the detection of bacteria in this size range to those capable of growth on microbiological media. Deficiencies associated with viable-count techniques have been reviewed previously (2-5). In addition to the difficulties associated with bacterial detection, treatment of biological particulate contamination has proven difficult. Bacterial penetration of 0.2 μm microporous membrane filters, which are the current industry standard, has been reported by a number of workers in the pharmaceutical industry (6-9). In addition, the effects of biocide treatments in these systems have been rather short lived, owing, in part, to the presence of recalcitrant bacterial biofilms (10).

The ability of bacteria to act as self-replicating entities distinguishes these contaminants from other, abiological, particulates. Their growth, replication, and production of various ionic and organic by-products in otherwise contaminant-free purified waters presents a tremendous challenge to production and quality assurance personnel. The same metabolic and structural properties which enable survival in such an otherwise hostile environment create special problems for bacterial detection and treatment.

This paper describes some of the design strategies which have been employed for limiting biological growth and biofilm formation. In addition emerging technologies for in situ detection of particulate and biofilm bacteria and their organic by-products are explored. The microbial growth cycle in purified water environments will be described as it relates to the formation of biological particles. Finally, some novel approaches to controlling biofilm populations in purified waters will be introduced.

BACTERIAL CONTAMINATION CASE HISTORIES

Pharmaceuticals and Medical Devices

A number of product recalls have been initiated for contaminated solutions marketed as sterile. Anderson (6) described contamination of sterile distilled water respiratory care solutions as a result of *Pseudomonas pickettii* passage through sterilizing-grade 0.2 μm membrane filters. Eye-care solutions have been contaminated by

Pseudomonas sp. originating in purified water feed systems (11). The origin of contaminated povidone iodine solutions by *Pseudomonas cepacia* was found to be a make-up purified water system (12). Novitsky et al. (13) described problems with removal of endotoxins emanating from the purified water system adsorbed to parenteral fluid containers.

Roberts (14) has described contamination of medical device sterilant solutions with *Mycobacterium chelonae.* Klein et al. (15) showed that 19% of the dialysate samples taken from acute dialysis centers in the U.S. were out of compliance with the Association for the Advancement of Medical Instrumentation bacteria standards. Murphy et al. (16) have described problems with endotoxemia during high-flux dialysis procedures.

Microelectronic Devices

Bacterial particulates have been implicated in a number of device failures in the U.S. and abroad. In one study performed at a Hewlett Packard facility in California, Dial and Chu (17) reported a direct correlation between purified water bacterial levels and device defect frequencies. Craven et al (18) have attributed failures in back-side metallization processes to bacterial adhesion processes. Eisenmann and Ebel (19) described horizontal and vertical defects built in during the photoresist steps resulting from contamination by bacteria and other particulate. Conductance, electromigration, and corrosion in oxide layers were also noted. Sodium and potassium ions leaching from bacteria during metabolism or cell death can diffuse into silicon wafer surfaces to change the device field effects.

Organic contamination resulting from bacterial activities may be one source of total oxidizable carbon (TOC). Carbon-containing particulates, such as bacteria, have the potential to act as dopants (20). Among other effects, high TOC levels create problems with photolithographic deposition processes and hazing of quartz furnace tubes.

Harned (21) presented scanning electron microscopic (SEM) evidence of bacterial contamination of MOS devices in a Motorola facility. Similar evidence of semiconductor device contamination was presented earlier by Crooke and Lutch (22). They surmised that bacterial contamination was responsible for alkali attack of "high packing density" semiconductor logic circuits. They found that high bacteria levels (>8 ml^{-1}) in deionized processing waters were the source of contamination on the IC devices. Conductance phenomena resulting from bridging of adjoining circuit paths by bacteria are a potentially devastating problem (Fig. 1).

In addition to the serious product contamination problems reviewed above, biological contamination of purified water system production and distribution systems can result in significant material and labor costs. Permanent losses in reverse osmosis flux and salt rejection, plugging of submicron membrane and ultrafilters, contamination of primary and polishing deionization resins, and fouling of piping and storage systems have all been attributed to bacterial growth and biofilm formation (2). Stoecker and Pope (23) have described corrosion processes in a high temperature DI water system resulting from microbially influenced corrosion activities.

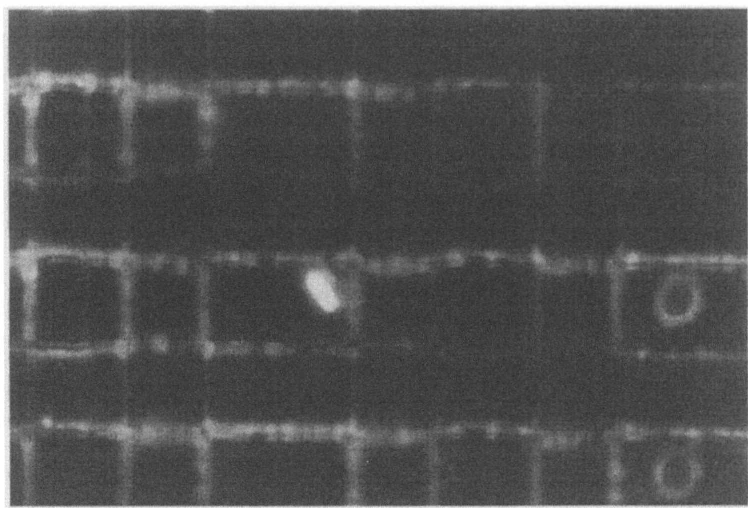

Fig. 1. Contamination of an LSI semiconductor device with 1.5 μm psuedomonads in 18 MOhmcm purified water

BACTERIAL GROWTH AND BIOFILM FORMATION

Virtually all bacteria isolated from purified water systems are Gram-negative rods. The gram designation refers to the bacterial cell wall structure. The Gram-negative cell wall consists of multiple layers of phospholipids surrounded by a lipid-polysaccharide structure (LPS). This multi-laminate structure may afford the cell protection from the extremely hypotonic environment which is intrinsic to purified water systems. These bacteria are heterotrophic in nature, requiring the presence of reduced organic compounds as energy sources. Therefore, these compounds serve as limiting growth factors in purified water systems. The term "oligotroph" has been assigned to organisms which are capable of growth in media containing <1 mg l^{-1} organic carbon (24). In general, a positive correlation exists between *assimilable* organic carbon levels and planktonic bacterial numbers in purified waters (Fig. 2).

In addition to organic energy and carbon sources, bacteria require a number of other nutrients such as nitrogen, phosphorus, sulfur, and trace metals and salts in order to carry out normal metabolic and replicative functions (2). A common-sense approach to controlling biological contamination would isolate and remove these growth-limiting factors from the system. Table 1 lists the major factors and some common reservoirs.

When one or more of these essential growth factors is limited, as is often the case in benthic environments and purified water systems, bacteria utilize a number of strategies designed to assure their survival. In the short term, "starved" bacteria tend to utilize endogenous energy reserves for replication; therefore, one strategy for survival involves replication to increase the probability of species survival once additional energy sources become available (25). In marine systems limited in assimilable nutrients, some bacteria reduce their cell volume, forming "ultramicrobacteria". Tabor et al. (26) demonstrated that these organisms, many of which were <0.3 μm in diameter, were capable of passing through 0.45 μm pore-sized microporous membrane filters. Their findings tend to corroborate the work of Christian and Meltzer (7) and others regarding bacterial penetration of "sterilizing-grade" microporous membrane filters.

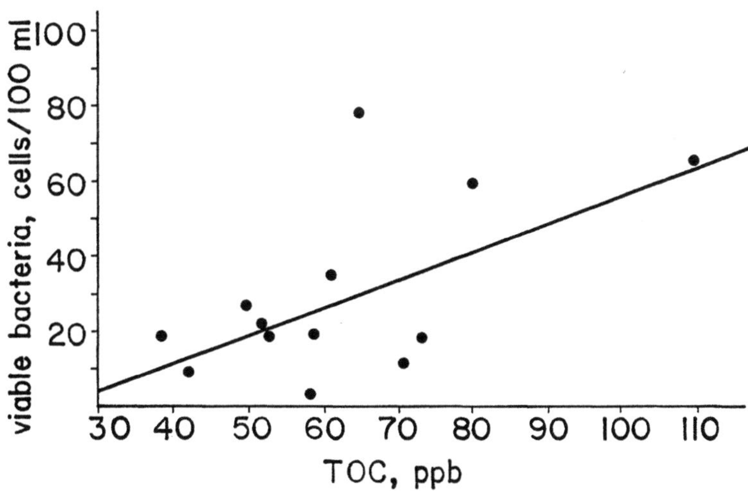

Fig. 2. Relationship between total oxidizable carbon (TOC) and viable bacteria numbers in a semiconductor purified water system. Data presented were obtained over 60 days

Perhaps the most significant adaptive mechanism utilized by nutrient-limited purified water bacteria involves adhesion to surfaces (Figure 3). Indeed, the majority of bacteria in nutrient-limited environments - such as an ultrapure water system - are attached to surfaces. Surfaces afford these organisms three major advantages over the bulk phase: 1) Concentration of trace organics, which can serve as nutrient sources, occurs on clean surfaces shortly after their immersion. 2) Surface-associated bacteria tend to produce extracellular polymeric substances which can further concentrate trace growth factors. 3) Bacteria in biofilms are afforded some protection from antagonistic agents such as biocides, heat treatments, and other inhibitory factors. Several workers have shown that nutrient-limiting environments promote the attachment of bacteria to surfaces (2,27).

Table 1. Growth-promoting factors in purified water systems.

Substrate	Source(s)
carbon	pipe extractables microbial by-products airborne dust treatment chemicals (RO pretreat) personnel (skin flakes, etc.) lubricants membrane filter surfactants storage tank entry hatch gaskets membrane filter media
water	vent filters degasifier HEPA filters storage tank sight tubes dead-legs
nitrogen	feed waters (humic & fulvic acids, nitrites, etc.) airborne dust nitrogen blankets microbial by-products
phosphorous/sulfur	feed waters (phosphates, sulfates) sulfuric acid (RO pretreatment) airborne dust membrane surfactants microbial by-products
trace metals and salts	airborne dust (degasifiers) piping extractables membrane filter extractables RO pretreatment chemicals metallic system components
light (algae/diatoms)	storage tank sight tubes entry hatch gaskets vent filters

Fig. 3. Transmission electron micrograph of bacteria within a polymeric biofilm. Bar = 1.0 μm

The relatively high surface area:volume ratio associated with industrial water systems provides ample space for bacterial attachment. Any one of the many systems utilized in the purification of feedwaters is therefore a potential reservoir for contamination. In terms of biological contaminant generation, however, granular activated carbon columns (28, 29), reverse osmosis membranes (30), ion exchange systems (31), degasifiers (2), RO/DI water storage tanks (32), and microporous membrane filters (33, 34) are the most significant areas of concern. Sources of contamination in industrial water systems have been reviewed by Mittelman and Geesey (2).

Bacterial biofilm formation has significant implications for both the treatment and detection of contaminating purified water populations. LeChevallier et al. (35) showed, for example, that attachment of Klebsiella pneumoniae to glass slides increased resistance to free chlorine disinfection by a factor of 150. It has been suggested that the presence of polyanionic extracellular polymeric substances affords cells some measure of protection by inhibiting biocide diffusion. Since biofilm-associated bacteria are not detected by current test methodologies, serious underestimates of contaminating populations result. Conversely, this inability to detect the major source of bulk phase biological contaminants results in an overestimate of treatment efficacies.

DESIGN AND PREVENTATIVE MAINTENANCE CONSIDERATIONS

Microporous Membrane Filters

In addition to increases in transmembrane pressure, membrane degradation, and pyrogen production, a major consequence of membrane biofouling is bacterial penetration or "grow-through" (Fig. 4).

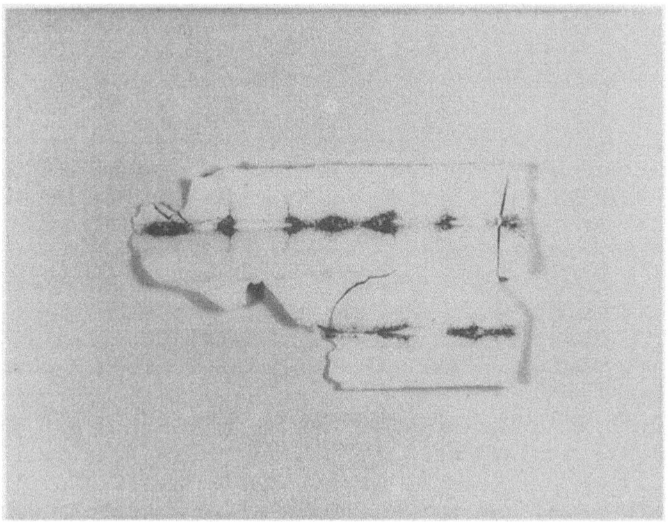

Fig. 4. Bacterial contamination of an 0.2 μm cellulose acetate microporous cartridge filter

Bacterial passage through microporous membrane filters has been described by several workers (7, 8). Bacterial penetration occurs under both dynamic and quiescent flow conditions. Bacteria have been shown to penetrate membranes whose pore size is smaller than that of the smallest cellular aspect (33). Membranes with relatively narrow pore-size distributions are also susceptible to grow-through (8). Wolf and Schoppmann (36) have described the penetration of 0.2 μm membranes (polycarbonate and cellulose nitrate) by streptomycetes. They suggest that passage is mediated by constriction of vegetative filaments rather than deformation of pores.

It is likely that the limiting factor affecting penetration is the thickness of bacterial cell walls and essential organelles. The aggregate size of these structures is in the range of 0.1 μm for many bacteria.

The current FDA (37), Health Industry Manufacturers Association (38), and American Society for Testing and Materials (39) requirements for viable particle retention are designed to evaluate filtration efficacy. The assay involves a challenge of 10^7 cells cm^{-2} of feed water surface. The challenge organism, *Pseudomonas diminuta* is grown in a defined saline lactose broth such that a fairly uniform population of 0.2 μm cells results. The cells are passed through the membrane on a one-time basis and a determination of viable particle passage is made using standard cultural techniques. Passage of less than 1 cell/10^7 cm^{-2} constitutes membrane failure.

Membrane fouling or grow-through potential are not, however, addressed by these standards. Perhaps of more concern to manufacturers is the apparent lack of insight into what factors might contribute to the development of conditions promoting grow-through. Despite deficiencies inherent to the current assays, it is important that microporous membranes be evaluated for retention efficacy prior to their installation. The large surface area and surface charge associated with pleated and disc-type membranes can provide a suitable substratum for bacterial biofilm development.

Materials of Construction and Hydraulic Effects

The pharmaceutical industry has long recognized the importance of purified water systems design and materials of construction in limiting bacterial growth. Dead leg areas, threaded fittings, and low-flow distribution sections are minimized or eliminated. While flow rates and surface inhomogeneities do appear to play some role in adhesion and biofouling processes, biofilms can and do form on smooth surfaces exposed to high shear environments. The advantages of higher flow rates and reduced dead-leg areas are in mitigating nutrient concentration polarization and increasing the frequency of ultraviolet and filtration systems contact.

Experiments in oligotrophic marine environments suggest that shear forces induced by increased flow rates may not be sufficient to prevent bacterial colonization and/or remove adhered cells. Mittelman et al. (40) showed that colonization of polished stainless steel surfaces by *Pseudomonas atlantica* was enhanced at higher fluid shear forces up to a critical shear of approximately 12 N m^{-2}. Characklis (41) has suggested that fluid hydraulics may affect the initial rate of colonization; however, following this initial "lag" phase, biofilm development appears to proceed somewhat independently of fluid flow.

The U.S. Navy has been evaluating various surfaces for their resistance to biofilm formation for over one hundred years. As yet, they have failed to develop materials or surface characteristics which prevent microbial adhesion. In addition, most of the biocidal coatings which have been employed are effective for relatively short periods of time (42). In any case, these coatings all have intrinsic release rates, which militate against their use in purified water systems.

One approach to limiting attachment processes would be to interfere with those physicochemical forces which mediate the initial adhesion event. Reviews of efforts in this direction have been made by Absolom (43), Baier (44), and Rosenberg and Kjelleberg (45). In general, hydrophobic surfaces (e.g. Teflon) promote bacterial adhesion in purified water. Conversely, the rate of attachment is lower for higher energy, hydrophilic surfaces (e.g. glass). However, once cells have colonized these two types of surfaces, they tend to desorb from the hydrophobic surfaces more rapidly; attachment to high energy surfaces appears to be less reversible.

In this regard, some preliminary work suggests that incorporating alternating hydrophobic/hydrophilic regions in polymers used in distribution system piping may retard bacterial adhesion. Clearly, more research is needed to further characterize those factors influencing both the initial stages of bacterial adhesion and the stability of the adhered cells and biomass.

Point-of-Use Distribution

Personnel contact with product water dispensing systems is a major source of contaminants. Attention should therefore be given to the most common types of system: sink goosenecks, hand-spray rinsers, hoses, cascading sinks, and dump rinsers. Contaminants such as skin oils, hairs, and respiratory droplets can serve as a direct source of bacteria as well as a source of growth-promoting substrates. The low flow and turnover rates associated with some of these dispensing systems can also create dead leg areas. Weekly treatments of dump rinsing and cascading sinks with a 10% (v/v) hydrogen peroxide solution for thirty minutes is an effective preventative treatment for these systems.

Several types of end-use treatment devices have been employed to control growth at the nozzle orifice. A 254 nm ultraviolet light has been used on the orifices of laboratory goosenecks to limit adventitious contamination from the outside environment. The system, which is no longer in production (EPCO Corporation, Minnesota), did appear to be effective is this regard. However, there is no evidence to suggest that external contamination of faucets by cleanroom air is a major source of bacteria. A microbial "barrier faucet" has been developed for the U.S. space transportation systems (shuttle) employing a laminar flow air-disinfectant flush system (46).

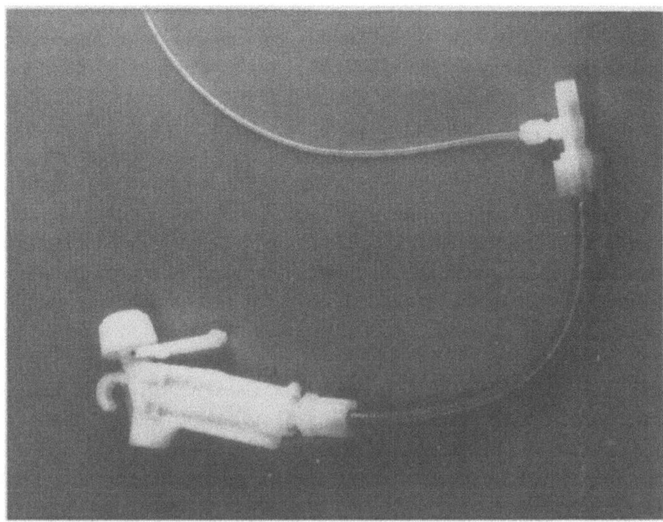

Fig. 5. Recirculating hand-spray rinser

Hand spray rinsers, by virtue of their discontinuous operation and large internal surface area, can be a significant dead-leg reservoir for bacteria. Fluoroware Corporation (Chaska, Minnesota) has developed a recirculating hand-sprayer (Fig. 5) designed to limit the effects of nutrient and bacterial polarization.

Ultraviolet Systems

Ultraviolet sterilization systems are typically employed for the control of planktonic bacterial populations within purified water systems. When these systems are properly installed and maintained, they provide greater than a 99.9999% reduction in *viable* bacterial numbers (47). The poor penetrating power of UV light renders this type of treatment relatively ineffective against bacteria in biofilms. Where bacteria (and other microorganisms) are inactivated by UV treatment, they are often removed by filtration. UV inactivated cells may then serve as nutrients for viable cells and/or a source of pyrogenic contaminants.

Newly installed UV lamps are usually designed to provide approximately 60,000 μWscm^{-2}; after one year of service, the lamps should provide a minimum of approximately 20,000 μWscm^{-2}. Lamps based on mercury/argon mixtures exhibit the highest efficiency. Following a one year service time, the lamps should be replaced and quartz sleeves cleaned. Many manufacturers recommend a quarterly inspection with a UV dosimeter to insure proper lamp operation. Since UV light efficacy is primarily a function of energy and contact time, it is important that the design flow rates not be exceeded.

A recent application of UV systems in purified waters has been in the reduction of organic carbon levels. Low pressure lamps with resonance wavelengths in the range of 185 nm are typically used in these applications. Aquafine Corporation in Valencia, California has taken a leading role in promoting this organic removal technology.

NOVEL DETECTION METHODS

Real-Time Monitoring

In any process, it is important to monitor possible contamination so corrective action can be instigated as rapidly as possible. The ideal monitors should be non-destructive so they will not inhibit or damage the biofilm. Thus the sensors will provide a true picture of the problem. If the biosensors can be placed in supply lines or in the system ahead of purification/disinfection systems then treatments can be modified to maintain the desired level of purity.

Deficiencies Associated with Cultural Techniques

The classical evaluation methods that rely on the culturing of organisms suffer from several serious deficiencies. Organisms from extreme environments (ultrapure water is an extreme environment) are notoriously difficult to culture so the standard plate counts will not give an accurate estimate. The problem of viable but non-culturable bacteria is a well accepted concept in public health microbiology. The organism causing cholera,

Vibrio cholera, when starved, forms minicells that are often not culturable but readily initiate the disease when the water is consumed.

Not only do the microorganisms not grow out readily on attempted isolation, but the growth of many oligotrophic organisms is extremely slow. This delay is clearly not acceptable for process waters in which continual control is essential.

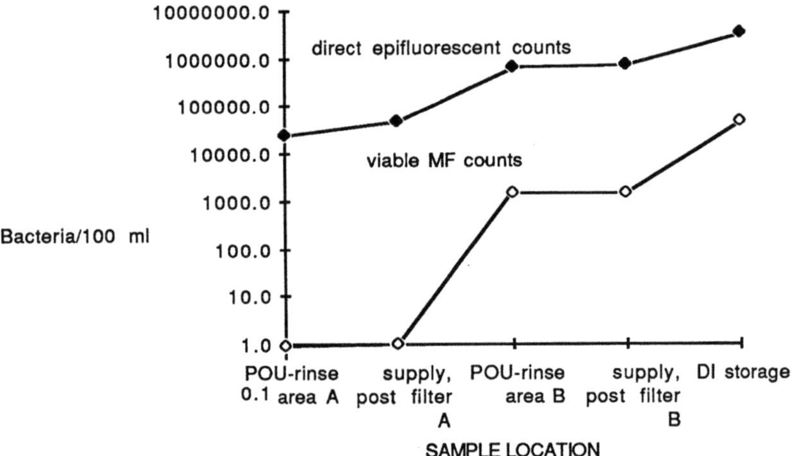

Fig. 6. Relationship between total direct and viable bacterial counts in an 18 Mohmcm water system

Several of the classical microscopic techniques are not very sensitive. They often require at least 10^3 organisms ml^{-1} in small samples for reliable detection. Concentration by centrifugation or filtration with vital staining and microscopic examination requires that the water system be sampled (thereby creating a potential contamination), is time consuming, and requires trained operators for monitoring. Usually, the type(s) of bacteria involved in the contamination are not readily determined with the microscopic examination. It is possible to identify specific microbes using fluorescent labeled probes. However, these techniques require that the organisms commonly contaminating the system be well known so the antibody or gene probes can be developed. Despite these deficiencies, the recoveries associated with direct counting techniques are often significantly greater than those obtained with cultural techniques (Fig. 6).

Bulk Phase Monitoring

Colloidal particles of bacteria in flow cells or test systems can be detected with laser light scattering. In some cases specific types of bacteria can be differentiated (48). A newly designed laser light scattering device with a flow cell (Wyatt Industries, Santa Barbara, CA) which utilizes 360° diode array detection greatly increases the resolution and sensitivity of this technique.

Resonance Raman Spectroscopy

Lasers which provide high power at frequencies between 190 and 260 nm can differentially activate components in the bacterial cell walls and cytoplasm and produce a shift in the wavelength of the scattered photons which can be detected as a resonance Raman spectrum (49). With a highly focused pulsed laser it is possible to detect a single cell. This group has shown selective excitement of bacterial cellular components using a tunable laser. By progressively shortening the incident wavelength, nucleic acids, calcium dipicolinate in endospores, tyrosine and tryptophane in the proteins, and the other aromatic amino acids can be distinguished by their vibronic fingerprint spectra. The patterns are specific for groups of bacteria and can be utilized for identification and can give information on the nutritional status of the bacteria (50) as they pass through UV transmissive capillaries.

In situ **biofilm monitoring**

In ultrapure water systems the majority of contaminating bacteria are found associated with surfaces where the sparse nutrients accumulate. On the surfaces they will form biofilms which are much more resistant to chemical countermeasures than the same organisms in the bulk fluids.

Fourier transforming infrared spectrometry (FT/IR) operated in the attenuated total reflectance mode allows the detection of bacterial biofilms as they form on a crystal of zinc selenide or germanium (51). The amide stretching of the proteins and ether stretch of the carbohydrates are clearly detectable when bacteria attach to surfaces. Often the nutrients attracted to the surface from the bulk phase are also clearly indicated by their infrared fingerprint. If the water system contained IR "windows", biofilm formation could be monitored and an indication of the chemical nature of the contamination would be apparent. Periodically, some mechanism for cleaning the "window" would need to be provided.

The technique also provides information on the nutritional status of the biofilm microbes as the formation of the endogenous storage polymer poly beta-hydroxyalkanoate (PHA) can be detected in the spectra. PHA forms during conditions of unbalanced growth when some critical shortage prevents bacterial cell division.

In the presence of precious metal microstructures, the electron plasmon can markedly potentiate the Raman scattering of molecules intimately associated with the surface in surface enhanced Raman spectroscopy (SERS). In theory, it should be possible to use this technique to non-destructively detect the attachment of bacteria or bacterial derived products to the metal microbases. The Raman fingerprint should be diagnostic of the chemical composition of the attaching species. Since this can be a cumulative effect, the sensitivity should be very great (White, unpublished data).

If electrically isolated electrodes can be incorporated into the system it is possible to monitor biofilm formation and persistence electrochemically. There are two types of electrochemical monitoring systems. In the first system, the potential between a standard electrode and a working electrode is monitored. The open circuit potential becomes more negative as the biofilm forms on the electrode surface (52). The potential shifts as metabolic activity in the biofilm changes.

The effects of the biofilm may also be monitored on the same coupon using a frequency spectrum of the noise (electrochemical noise). Perturbation-response methods may also be utilized to show changes in the surface chemistry as a result of biofilm activity.

In electrochemical impedance spectroscopy, a small sinusoidal potential is induced in the system over a 5 decade frequency range and the induced changes in current and phase shift angle determined.

Small amplitude cyclic voltammetry can also be used to monitor biofilm activity as it changes the electrochemical activity on the working electrode surface. These methods are reviewed relative to microbially influenced corrosion in Dowling et al. (53).

The presence of material of a significantly different intrinsic viscosity than that of the water can be detected as a change in frequency of a vibrating quartz crystal in quartz crystal microbalance (QCM) technology. The surface can also be used as an electrode to produce both the microbalance and electrochemical signals (54). This holds a potential for a non- destructive biofouling monitor.

Bacterial pyrogens (endotoxins)

Significant insight into the nature of contamination of ultrapure water can be gained from destructive analysis of biofilms or filter retentates recovered from the system. The most important contaminants of pure water systems are gram negative bacteria or the lipopolysaccharide (LPS) products from these organisms. Novitsky (4, 55) has reviewed the significance and detection of endotoxins in semiconductor-grade and other ultrapure waters.

The lipid A of the LPS of gram negative bacteria evokes a most powerful physiological response when injected into human beings. The mildest response is fever (pyrogen) which can progress to anaphylactic shock and death. The massive inflammatory response provoked by endotoxin represents the human bodies natural response to the danger of gram negative septicemia. Consequently it is of paramount importance that water for intravenous infusion and medical devices for implantation be free of gram negative bacteria or their endotoxic products.

Currently, pyrogens are tested by inducing a biological response with the *Limulus amebocyte* lysate (LAL) test. This test allows detection of as few as approximately 10^3 cell equivalents of endotoxin. The test is a bioassay that depends on the activation of proteins by the endotoxin. It is not effective in solutions containing alcohol, chelating agents, urea, high salts or some proteins (55). This test is most effective against endotoxin or bacteria recovered from the water as the endotoxin attaches to surfaces quite readily.

Novitsky (55) found that a significant positive correlation exists between free and free and bound endotoxin and bulk-phase bacterial concentrations in ultrapure water. The methods described are rapid (requiring approximately 1 h incubation) and amenable to automation. The extrapolated detection limit for bacteria in ultrapure water is on the order of 100-1000 cells ml^{-1}.

A new endotoxin assay that is bacterial family/species specific, more sensitive, and less subject to interferences has recently been developed (56). The test will also detect endotoxin attached to surfaces. The technique depends on the extraction of lipids from the biofilms or filter retentates, hydrolysis of the lipid-extracted residue, and re-extraction with recovery of the previously ester- or amide-linked hydroxy fatty acids in the lipid A of the LPS (57). Derivatization of the carbonyl and hydroxyl moieties allows the separation of the different hydroxy fatty acids by capillary gas chromatography and detection by chemical ionization mass spectrometry of the negative ions provides sensitivity to the equivalency of a few cells. The composition patterns of the different LPS hydroxy fatty acids provides identification of bacterial families. Recent advances have led to the possibility of automating this analysis.

Phospholipid ester-linked fatty acid (PLFA) analysis of biofilms or filter retentates can provide definitive identification of contaminating organisms (58, 59). The presence of phospholipids indicates the presence of viable or potentially viable cells as the endogenous phospholipase activity rapidly degrades any polar lipids in non-viable cells. Thus, phospholipids are an accurate measure of the cellular biomass. The specificity of PLFA compositional patterns to specific subsets of the microbial community allows quantitative documentation of the microbial community structure. Analysis of other lipid components of the system provides insight into the community nutritional structure that can be important in testing countermeasure efficacy. The metabolic activity of subsets of specific components of the community can be readily determined by analysis of ^{13}C isotope enrichment after exposure to mass labeled precursors with the great specificity and sensitivity of gas chromatography/mass spectrometry.

Treatment considerations

Influence of Biofilms on Treatment Efficacy

The so-called "bacterial regrowth" phenomenon following chlorine treatments of purified and potable water systems is likely a function of biofilm formation rather than resistance in the classical sense of antibiotic resistance (60, 61). A combination of the polyanionic nature of many EPS moieties coupled with the inherent biocide demand associated with extracellular slime matrices acts as an inhibitor of biocide actions. The fact that most power plant condenser waters are treated on a continuous or frequent slug dosage basis is an indication of the recalcitrance of fouling biofilms. A treatment strategy which focused on modifications of the primary conditioning film and/or initial colonizing bacterial population might obviate the need for toxic biocides, thus removing a significant threat to aquatic ecosystems.

Inactivation and removal of bulk phase bacteria following the most commonly employed chemical and physical treatments is usually complete. Within days or weeks of treatment, however, bulk phase bacteria levels are often as high or higher than pretreatment levels. Bacteria existing in sessile biofilms are protected from the effects of biocides and, to some extent, physical treatments. Numerous studies have shown that repopulation of purified and other industrial water systems is mediated by the presence of adherent bacterial populations. The key to effective treatments, therefore, is inactivation and removal of this population of contaminating biological particulates.

Biocide (Chemical) Treatments

The development of process and product compatible biofouling treatments has long evaded manufacturers of critical products which require purified water as a raw material

(62). Of all the physical and biocide treatments applied to semiconductor-grade purified water systems, ozone appears to hold the greatest promise as an effective, product compatible biocide. The technology is currently available for inactivation of residual ozone levels in situ via UV irradiation (63). Most of the problems with ozone application to purified water systems have been in the realm of component compatibility. Since ozone is an extremely reactive oxidant, its use is confined to systems constructed of such inert materials as Teflon and other fluoropolymers. Table 2 lists some of the more commonly applied chemical biocides along with typical dosage regimes.

Table 2. An overview of purified water biocide treatments

Treatment	Dosage regime	Treatment duration
quaternary ammonium compounds	300-1000 ppm	2-3 h
chlorine	300-1000 ppm	2-3 h
peracetic acid/peroxide	1% (v/v)	2 h
iodine	50-100 ppm	1-2 h
hydrogen peroxide	10% (v/v)	2-3 h
formaldehyde	1% (v/v)	2-3 h

Hydrogen peroxide, a commonly employed purified water system biocide, can also be readily degraded by ultraviolet light. However, it has relatively little activity against attached bacteria in biofilms. There is also evidence which suggests that hydrogen peroxide requires catalytic concentrations of Fe^{2+}, Ni^{2+}, or Cu^{2+} for optimal biocide efficacy (64).

Surfactants such as the quaternary ammonium compounds have excellent biocide activity in addition to their intrinsic detergency. Synergistic combinations of chlorinated or brominated compounds with surfactants may provide additional activity against bacterial biofilm populations. Their ability to interact with surfaces does, however, create problems related to removal of these compounds following application. Larger volumes of rinse water are required for the surfactant compounds, than, for example, hydrogen peroxide.

Physical Treatments

The use of high temperature water or steam flushes for biofilm removal has proven effective in pharmaceutical-grade purified water systems. Hot water maintained at or above 80 °C will prevent the growth of all bacteria encountered in a purified water system. Purified water-for-injection (WFI) used in the production of injectable drugs is typically stored and recirculated through stainless steel storage tanks which are maintained at a constant temperature of 80 °C. Flowing steam is often employed in the biotechnology industry to sterilize stainless steel fermentation vessels and attendant piping. So-called sterilize-in-place (SIP) systems employing hot water or steam have been used for many years in the dairy and food products industries. The utility and economy of such systems in the semiconductor industry is, however, questionable.

A patented process developed by J. W. Costerton (65) utilizes a circulating glycol solution to effect ice nucleation within the biofilm. Under controlled thawing conditions, the frozen cells and biomass are stripped from the surface and trapped via side-stream filtration devices. The applicability of this process to a broad industrial market is, however, still to be validated.

Future treatment strategies will undoubtedly focus on the prevention of primary bacterial colonization. This will entail alterations in the surface chemistry of both the bulk fluid and the substratum. Although great progress has been made in developing advanced fluid handling materials of construction, mankind still has not developed a surface to which bacteria will not adhere. There exists a real need for process and product compatible biofilm treatments.

ACKNOWLEDGEMENTS

Some of the research reported in this paper was presented at the Ninth Annual Semiconductor Pure Water Conference, January 17-18, 1990, in Santa Clara, California. The author's research is supported, in part, by a grant from the U.S. Office of Naval Research, contract number N00014-87-K-0012 and the National Aeronautics and Space Administration, contract number NAS8-38493.

REFERENCES

1   Yang M, Tolliver DL (1989) Ultrapure water particle monitoring for advanced semiconductor manufacturing. J. Environ. Sci. July/August, 35-42

2   Mittelman MW, Geesey GG (1987) Biological fouling of industrial water systems: a problem solving approach. Water Micro Associates, San Diego

3   Morita RY (1985) Starvation and miniaturization of heterotrophs, with special emphasis on maintenance of the starved viable state. In: Fletcher M, Floodgate G (eds) Bacteria in the natural environments: the effect of nutrient conditions, Society for General Microbiology, London; 111-130

4   Novitsky TJ (1984) Monitoring and validation of high purity water systems with the Limulus amebocyte lysate test for pyrogens. Pharm. Engin. 4, 21-33

5   Reasoner DJ, Geldreich EE (1985) New medium for the enumeration and subculture of bacteria from potable water. Appl. Environ. Microbiol. 49, 1-7

150

6   Anderson RL, Bland LA, Favero MS, McNeil MM, Davis BJ, Mackel DC, Gravelle CR (1985) Factors associated with *Pseudomonas pickettii* intrinsic contamination of commercial respiratory solutions marketed as sterile. Appl. Environ. Microbiol. 50, 1343-1348

7   Christian DA, Meltzer TH (1986) The penetration of membranes by organism growthough and its related problems. Ultrapure Water 3, 39-44

8   Howard G, Duberstein R (1980) A case of penetration of 0.2 μm-rated membrane filters by bacteria. J. Parent. Drug Assoc. 34, 95-102

9   Simonetti JA, Schroeder HG (1984) Evaluation of bacterial grow-through. J. Environ. Sci. 27, 27-32

10   Mittelman MW (1986) Biological fouling of purified water systems: part III, Treatment. Microcontamination 4(1); 30-40; 70

11   Wilson LA, Schlitzer, Ahearn DG (1981) *Pseudomonas* corneal ulcers associated with soft contact lens wear. Am. J. Ophthalmol. 92, 546-554

12   Michels DL (1981) Validation and control of deionized water systems. FDA Abstracts August.

13   Novitsky TJ, Schmidt-Gengenbach J, Remillard JF (1986) Factors affecting recovery of endotoxin adsorbed to container surfaces. J. Parent. Sci. Technol. 40, 284-286

14   Roberts C (1988) Direct detection and enumeration of Mycobacteria in disinfectants by epifluorescence microscopy. Abs. Ann. Meet. Amer. Soc. Microbiol., May 8-13, Miami Beach, Florida.

15   Klein E, Pass Ted, Harding GB, Wright R, Million C (1990) Microbial and endotoxin contamination of water and dialysate in the central United States. Artificial Organs 14(2), 85-94

16   Murphy JJ, Bland LA, Davis BJ, Maxey RW, Light A, Favero MS, Solomon SL (1987) Pyrogenic reactions associated with high-flux hemodialysis. ICAAC Abstract 27:109

17   Dial F , Chu T (1987) The effect of high bacteria levels with low TOC levels on bipolar transistors: a case study. Proceedings of the 67th annual Semiconductor Pure Water Conference, January, San Jose, CA; 178-193

18   Craven RA, Ackerman AJ, , Tremont PL (1986) High purity water technology for silicon wafer cleaning in VLSI production. Microcontamination 4(11), 1421

19 Eisenmann DE, Ebel CJ (1988) Sulfuric acid and DI point of use particle counts and resultant silicon wafer FM levels. Proceedings of the 9th annual meeting of the Institute For Environmental Sciences (ICCS), September 26-30, Los Angeles; 547-559

20 Poirier SJ (1985) The new role of TOC analysis in pure water system management. Proceedings of the 4th annual Semiconductor Pure Water Conference, January, San Francisco; 197-210

21 Harned W (1986) Bacteria as a particle source in wafer processing equipment. J. Environ. Sci. 24(3), 33

22 Crooke M, Lutsch AGK (1976) Process evaluation of high packing density semiconductor logic circuits by a scanning electron microscope. Proc. Electron Microscopy Soc. Southern Africa ann. conf., December, Johannesburg

23 Stoecker JG, Pope DH (1986) Study of biological corrosion in high temperature demineralized water. Paper no. 126, Proc. Nat. Assoc. Corr. Engineers ann. meeting, NACE Publications, Houston, TX

24 Ishida Y, Kadota H (1981) Growth patterns and substrate requirements of naturally occurring obligate oligotrophs. Microb. Ecol. 7, 123-130

25 Novitsky JA, Morita RY (1978) Possible strategy for the survival of marine bacteria under starvation conditions. Mar. Biol. 48, 289-295

26 Tabor PS, Ohwada K, , Colwell RR (1981) Filterable marine bacteria found in the deep sea: distribution, taxonomy, and response to starvation. Microb. Ecol. 7, 67-83

27 Marshall KC (1988) Adhesion and growth of bacteria at surfaces in oligotrophic environments. Can. J. Microbiol. 34, 503-506

28 Johnston PR , Burt SC (1976) Bacterial growth in charcoal filters. Filtr. Sep. 13, 240-244

29 Collentro WV (1986) Pretreatment part II: activated carbon filtration. Ultrapure Water 3(3), 39-44

30 Ridgway HF, Rogers DM, , Argo DG (1986) Effect of surfactants on the adhesion of mycobacteria to reverse osmosis membranes. Proc. semiconductor pure water conf., January, San Francisco; 133-164

31 Flemming HC (1987) Microbial growth on ion exchangers. Wat. Res. 21, 745-756

32 Youngberg DA (1985) Sterilizing storage tanks in a pure water system. Ultrapure Water 2(4), 45

33 Meltzer TH (1987) Filtration in the Pharmaceutical Industry. Marcel Dekker, New York, 493-544

34 Mittelman MW , White DC (1989) The role of biofilms in bacterial penetration of microporous membranes. Proc. Pharm. Technol. meetings, September 18-20, Philadelphia; 211-221

35 LeChevallier MW, Cawthon CD, Lee RG (1988) Factors promoting survival of bacteria in chlorinated water supplies. Appl. Environ. Microbiol. 54, 649-654

36 Wolf H, Schoppmann H (1989) Streptomycetes can grow through small filter capillaries. FEMS Microbiol. Lett. 57, 259-264

37 Code of Federal Regulations (1987) Guidelines on sterile drug products produced by aseptic processing. 21 CFR 10.90, June.

38 Health Industry Manufacturers Association (1982) Microbiological evaluation of filters for sterilizing liquids. HIMA 3(4), Washington, DC

39 American Society for Testing and Materials (1983) Bacterial retention of membranes utilized for liquid filtration. ASTM F838-83, Philadelphia

40 Mittelman MW, Nivens DE, Low C, White DC (1990) Differential adhesion, activity, and carbohydrate:protein ratios of *Pseudomonas atlantica* monocultures attaching to stainless steel in a linear shear gradient. Microb. Ecol. 19, 269-278

41 Characklis WG (1990) Biofilm processes. In: Characklis WG, Marshall KC (eds) Biofilms. John Wiley, New York; 195-231

42 Goldberg ED (1986) TBT, an environmental dilemma. Environment 28(8), 17-20

43 Absolom DR, Lamberti FV, Policova Z, Zingg W, van Oss C, , Neumann AW (1983) Surface thermodynamics of bacterial adhesion. Appl. Environ. Microbiol. 46, 90-97

44 Baier RE, Shafrin EG, Zisman WA (1968) Adhesion: mechanisms that assist or impede it. Science 162, 1360-1363

45 Rosenberg M, Kjelleberg S (1987) Hydrophobic interactions: role in bacterial adhesion. In: Marshall KC (ed), Advances in microbial ecology, vol 9, Plenum Press, New York; 353-393

46 Sauer RL (1981) The potable water. NASA STS-1 Technical Report No. N82-15711 06-51, 63-66

47 Powitz RW, Hunter J (1985) Design and performance of single-lamp, high-flow ultraviolet disinfectors. Ultrapure Water 2(1), 32-34

48 Wyatt PJ (1973) Differential light scattering techniques for microbiology. In: Norris JR, Ribbons DW (eds) Methods in microbiology. Vol 8, Academic Press, New York; 183

49 Manoharan R, E. Ghiamati E, Dalterio RA, Britton KA, Nelson WH (1990) Ultraviolet resonance raman spectra of bacteria, bacterial spores, protoplasts, and calcium dipicolinate. J. Microb. Meth. (in press)

50 Dalterio RA, Nelson WH, Britt D, Sperry JF (1987) An ultraviolet (242 nm excitation) resonance Raman study of live bacterial components. Appl. Spectrosc. 41, 417-422

51 Nichols PD, Henson JM, Guckert JB, Nivens DE, White DC (1985) Fourier transform-infrared spectroscopic methods for microbial ecology: analysis of bacteria, bacteria-polymer mixtures, and biofilms. J. Microb. Meth. 4, 79-94

52 White, DC, Jack RF, Dowling NJE, Franklin MJ, Nivens DE, Brooks S, Mittelman MW, Vass AA, Isaacs HS (1990) Microbially influenced corrosion of carbon steel. Proc. Nat. Assoc. Corros. Engineers ann. meet., April, Las Vegas

53 Dowling NJE, Stansbury EE, White DC, Borenstein SW, Danko JC (1989) On-line electrochemical monitoring of microbially induced corrosion. In: Kucubam GJ (ed) Microbial corrosion: 1988 workshop proceedings, Electric Power Research Institute, Palo Alto, CA (report EPRI R-6345;8000-26); 5-17

54 Deakin MR, Buttry DA (1989) Electrochemical applications of the quartz crystal microbalance. Anal. Chem. 61, 1147-1154

55 Novitsky TJ (1987) Bacterial endotoxins (pyrogens) in purified waters. In: Mittelman MW, Geesey GG (eds) Biological fouling of industrial water systems: a problem solving approach. Water Micro Associates, San Diego; 77-96

56 White DC, Mittelman MW (1989) Detection of endotoxins of gram negative bacteria, United States Patent Application (pending)

57 Parker JH, Smith GA, Fredrickson HL, Vestal JR, White DC (1982) Sensitive assay, based on hydroxy-fatty acids from lipopolysaccharide lipid A for gram negative bacteria in sediments. Appl. Environ. Microbiol. 44, 1170-1177

58 White DC (1986) Assessment of marine biofouling formation, succession, and metabolic activity. In: Thompson, MF et al (eds) Marine biodeterioration, advanced techniques applicable to the Indian Ocean, Oxford and IBH Publishing, New Delhi

59   White DC (1988) Validation of quantitative analysis for microbial biomass, community structure, and metabolic activity. Adv. Limnol. 31. 1-18

60   Ridgway HF, Justice CA, Whittaker C, Argo DG, Olsen BH (1984) Biofilm fouling of RO membranes - its nature and effect on treatment of water for reuse. J. Am. Water Works Assoc. 76, 94-102

61   Wolfe RL, Ward NR, Olsen BH (1988) Inorganic chloramines as drinking water disinfectants: a review. J. Am. Water. Works Assoc. 76(5), 74-88

62   Mittelman MW (1986) Trends in the detection and control of biological fouling in purified water systems. Ultrapure Water 3(6), 22-23

63   Nebel C, Nebel T (1984) Ozone: the process water sterilant. Pharm. Manufact. 1(2), 16-22

64   Block SS (1983) Disinfection, sterilization, and preservation, Lea & Febiger, Philadelphia

65   Costerton JW (1983) United States Patent No. 4,419,248

WHAT IS BIOCORROSION?

Gill Geesey
Department of Microbiology
California State University
Long Beach, CA 90840 USA

ABSTRACT

Microorganisms growing on surfaces perform a variety of metabolic reactions, the products of which may promote the deterioration of the underlying substratum. These reactions refer to biocorrosion when the substratum consists of a metal or metal alloy. The effect of corrosive microbial products on an underlying metal surface is exacerbated when their concentrations are permitted to increase to high levels as may occur when the microorganisms grow on the surface in a biofilm. The biofilm contains exopolymers which impede the diffusion of solutes and gases between the surface and the bulk aqueous phase. The biofilm also permits the development of highly structured microbial communities on the surface. The various species are able to collectively carry out metabolic activities that are potentially more corrosive to the underlying surface than could be achieved by a single species acting alone. These features of sessile microbial growth represent important prerequisites of biocorrosion.

INTRODUCTION

Corrosion is viewed primarily as a series of electrochemical reactions at a metal surface in contact with an electrolytecontaining aqueous phase. Until recently, biological processes were considered to contribute to relatively few forms of corrosion. Microbiologically-influenced corrosion (MIC) has been considered mainly in anaerobic environments where sulfide-producing bacteria are active. Von Wolzogen Kuhr and van der Vlugt (1) proposed an electrochemical process caused by sulfate-reducing bacteria (SRB). Since then other types of microorganisms, including aerobes, have been implicated in MIC. Whereas, it is generally agreed that MIC does not involve any new form of corrosion process, the number of different processes in which they participate appears to be greater than was originally believed. The following discussion highlights some of the more important microbiological processes that are believed to influence metal corrosion

HETEROGENEITY CAUSED BY MICROBIAL COLONIZATION OF SURFACE

Differential Aeration Cells

Corrosion may occur on a submerged metal surface as a result of uneven colonization of the surface by microorganisms. When a clean surface is submerged in an aqueous environment it is rapidly colonized by microorganisms present in the bulk

H.-C. Flemming · G. G. Geesey (Eds.)
Biofouling and Biocorrosion in Industrial Water Systems
Proceedings of the International Workshop on
Industrial Biofouling and Biocorrosion, Stuttgart, Sept. 13-14,1990

aqueous phase. As the surface-associated microbes replicate, microcolonies of each species are formed. The microcolonies are distributed in an uneven manner with some areas of the surface being less heavily colonized and exposed to the bulk aqueous phase than other areas. In environments where the bulk aqueous phase is aerated, the oxygen consuming activities of surface-associated bacteria can create an oxygen concentration gradient near the metal surface; the higher concentrations occurring in that area of the colony in contact with the bulk aqueous phase and the lowest concentrations developing at the bottom of the colony in contact with the metal surface (Fig. 1).

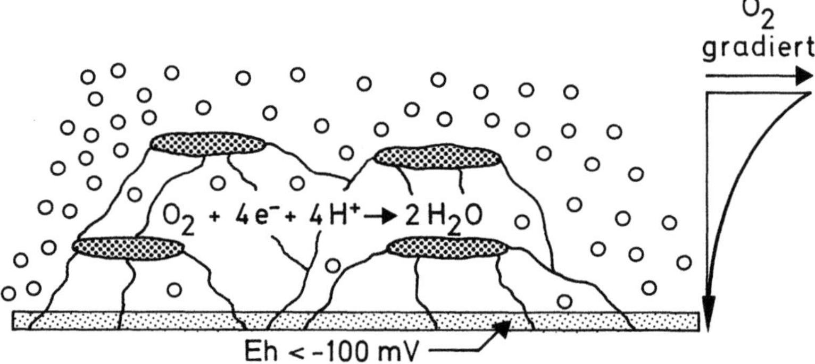

Fig. 1. Oxygen concentration gradient in biofilm caused by respiratory activity of microorganisms

An oxygen concentration cell is likely to develop where an uncolonized area of the surface in contact with the oxygenated bulk aqueous phase meets an area covered by a colony of oxygenrespiring bacteria; the area under the microcolony being anodic to the area exposed to the bulk aqueous phase (Fig. 2).

Differential Aeration Cell

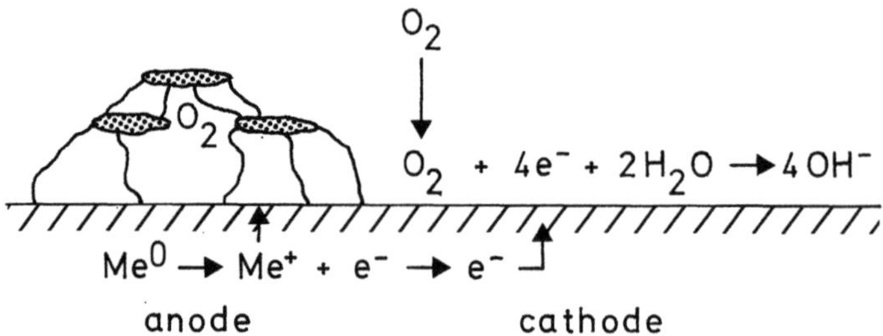

Fig. 2. Differential aeration cell resulting from heterogeneous distribution of biofilm-forming microorganisms over the surface

Little et al. (2) presented evidence that growth of a thermophilic bacterium on a 201 nickel brazed joint resulted in the establishment of a differential aeration cell at 60 °C but not at 23 °C; the higher temperature supported the growth of the bacterium whereas the lower temperature did not. Microorganisms also facilitate the formation of differential aeration cells on metal surfaces that contain an uneven distribution of corrosion product formed through abiological reactions. The precipitated metal salts provide sites for microbial attachment and colonization. Respiration by these bacteria further reduces the concentrations of oxygen under the deposit, the diffusion of which is already hindered by the presence of the deposit The reduced concentrations of oxygen under mineral deposits and microcolonies containing respiring microorganisms gives rise to localized areas that exhibit low redox potentials close to the metal surface. Such conditions permit the growth of facultative and obligate anaerobic bacteria even when the bulk aqueous phase is well-aerated. Thus, anaerobic sulfur-reducing and acidproducing bacteria, two groups of microbes that have been associated with corrosion, can grow in systems containing oxygenated water.

Microbial Consortia in Biofilms

Within a short period of time of exposure to most sources of water, virtually all metal surfaces become colonized by a variety of physiologically-distinct microbial species, the composition of which is determined by the surrounding environmental conditions. Microcolonies of different species develop next to one another and eventually merge to form a biofilm. Biofilms are a common mode of microbial existence in aquatic environments. They reflect a complex community structure within which diverse microbial activities take place. The biofilm restricts the diffusion of products of microbial metabolism excreted from the cells in each microcolony. Thus, very high concentrations of metabolic by-products may accumulate at or near the underlying metal surface. In instances where the surface is not compatible with the metabolites, corrosion may result.

Sulfur-reducing bacteria

Sulfur-reducing bacteria are highly dependent on the activities of other members of the microbial biofilm community (3). As obligate anaerobes, they depend on the oxygen respiring heterotrophs to reduce the oxygen concentration in the biofilm to low levels. They also depend on fermentative facultative anaerobes to supply organic electron donors for energy production. It is now known that sulfur reducing bacteria, as a group, can utilize a wide range of organic electron donors which include fumarate, acetate, propionate and fatty acids. The ability to utilize these organic compounds as well as hydrogen as electron donors contributes to the ubiquitous nature of this diverse physiological group of bacteria Sulfur-reducing bacteria may promote metal corrosion in several ways. Through the action of an active hydrogenase enzyme, these bacteria can prevent the accumulation of atomic or molecular hydrogen at the cathode, thus impeding cathodic polarization (1) (Fig.3). This enzyme activity alone, however, does not appear to be sufficient to account for the high rates of corrosion experienced in the field. Another theory considers the influence of the hydrogen sulfide produced during respiration of sulfate and other oxidized forms of sulfur by this group of bacteria. The sulfides of iron and hydrogen are effective cathodic depolarizing agents (4) (Fig.4). Sulfide also

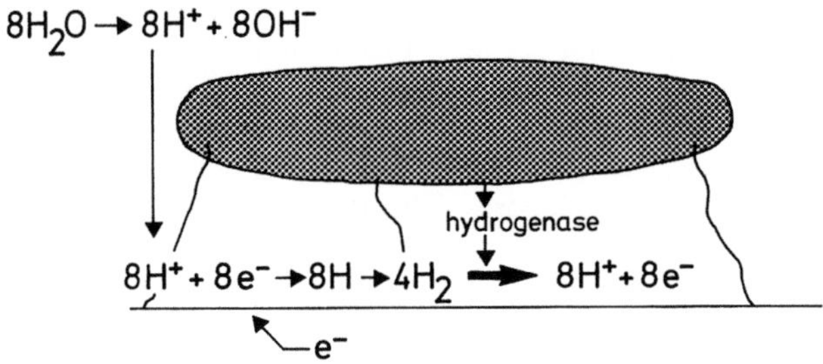

Fig. 3. Cathodic depolarization of surface due to utilization of hydrogen by hydrogenase-producing microorganisms

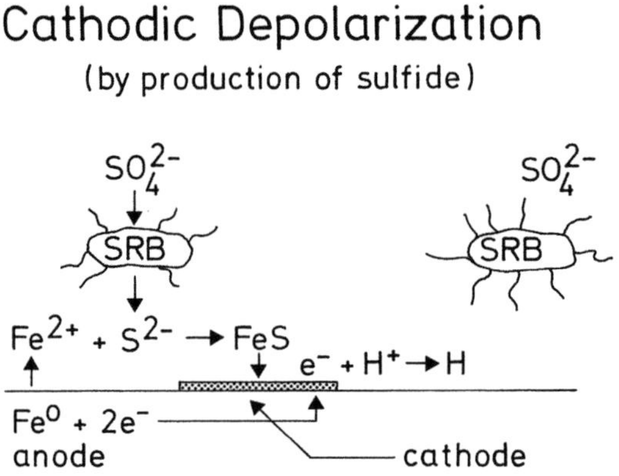

Cathodic Depolarization
(by production of sulfide)

Fig. 4. Cathodic depolarization of surface by iron sulfide as a result of respiration of sulfate by SRB's

$$H_2O \longrightarrow H^+ + OH^- \qquad\qquad HS^- \ (SRB)$$
$$2H^+ + 2e^- \longrightarrow 2H \not\longrightarrow H_2$$
$$RCOOH \longrightarrow H^+ + RCOO^-$$
$$HHHHHHH$$

(Metal surface)   **H**

Fig. 5a. Hydrogen embrittlement caused by sulfide generation by SRBs

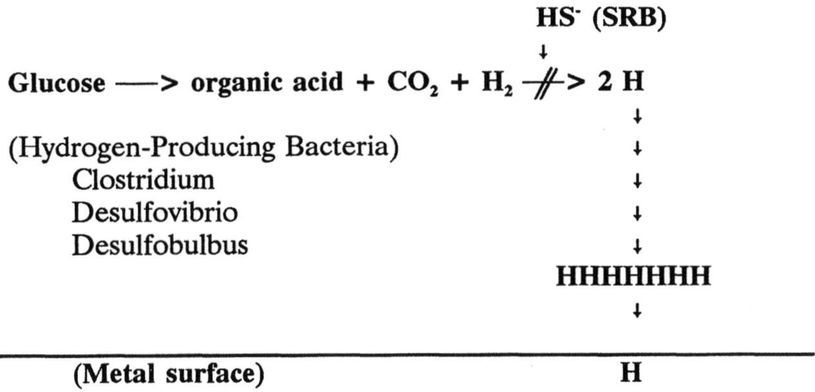

Fig. 5b. Hydrogen embrittlement caused by hydrogen-producing bacteria

Anodic Depolarization

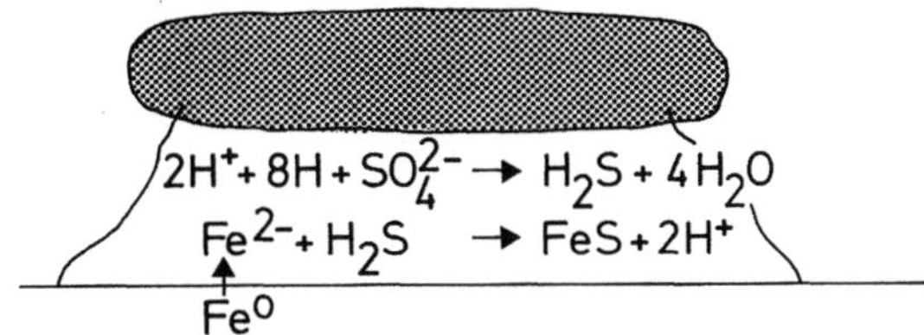

Fig. 6. Anodic depolarization of metal surface caused by reaction of hydrogen sulfide produced by SRBs

In the presence of sulfide, atomic hydrogen produced at the cathode as a result of the reduction of protons derived from the dissociation of water or acids may accumulate at the metal surface. This accumulation may facilitate adsorption of hydrogen into the metal, which in turn, may cause hydrogen embrittlememt in sensitive metals. Sulfide also promotes ionization of many metals and therefore accelerates the anodic reaction (Fig. 6). The relative contribution of sulfide formation and hydrogenase activity to a particular corrosion event likely depends on the nature of the metal surface and the conditions in the system.

Hydrogen-producing bacteria

Many microorganisms produce hydrogen gas as a product of carbohydrate fermentation (Fig. 5). Walch et al (5) showed that pure cultures of Rumenococcus albus produced small amounts of hydrogen that diffused into steel. It now seems that sulfide may reduce hydrogen adsorption by preventing the conversion of molecular hydrogen to atomic hydrogen. Whether the reactions resulting from such a microbial consortium are responsible for hydrogen embrittlement observed in the field remains to be verified.

Acid-producing bacteria

Some bacteria produce and excrete copious amounts of acid as by-products of metabolism. When the rate of production is greater than the rate of utilization by other microbial populations in the biofilm the acids may accumulate and create localized gradients in pH at the metal surface. Acid products of microbial metabolism cause thermodynamic "depolarization" of the cathode under deaerated conditions because the reversible potential for the hydrogen reaction shifts in a more noble direction with decreasing pH (Fig. 7).

Acid Production

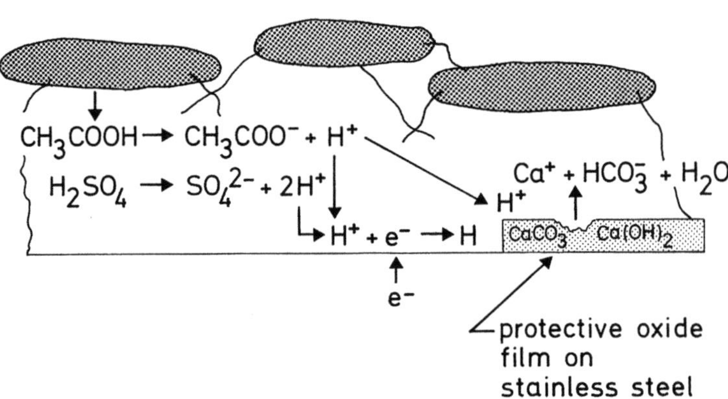

Fig. 7. Acid production (organic and inorganic) by adherent film-forming bacteria with consequent promotion of electron removal from cathode by hydrogen or dissolution of protective calcarious film on stainless steel surface

Acid production by the fungus *Cladosporium resinae* is believed to be the causes of corrosion of aluminum fuel tanks of jets. Little et al (2) suggested that the production of isobutyric and isovaleric acids by a thermophilic bacterium contributed to the corrosion of a 201 nickel brazed joint at elevated temperatures. In another study, Little et al (6) demonstrated that an aerobic acetic acid-producing bacterium (*Acetobacter aceti*) depolarized an Allegheny-Ludlum 6X stainless steel electrode cathodically polarized to -900 mV vs. a standard calomel electrode. The acetic acid produced by the bacteria destabilized or dissolved the protective calcareous film that was formed during cathodic polarization. Sterile solutions of similar composition to those which supported the growth of the bacteria promoted corrosion rates that were significantly less than those observed in solutions containing the bacteria Acid-induced corrosion of concrete and iron reinforcing bar used in the construction of sewage pipes can occur through the activities of autotrophic bacteria. *Thiobacillus thiooxidans* and *Thiobacillus ferrooxidans* produce sulfuric acid during metabolism of reduced sulfur compounds present in the sewage effluent. In some instances, the reduced sulfur is generated by sulfur reducing bacteria in the system.

Iron bacteria

Iron reducing bacteria can modify the protective oxide film that forms over a mild steel surface causing the surface to depolarize (7) (Fig. 8). Facultative anaerobes of Pseudomonas capable of using ferric ions and sulfite as terminal electron acceptors for anaerobic respiration use low molecular weight compounds such as lactate as carbon source. Some of these bacteria attach to mild steel coupons and remove a passive gamma Fe_2O_3-film and replace it with a biofilm under which pits developed. One isolate depolarized the anode (7). The reaction at the cathode was unaffected by the bacterium. The bacteria used the ferric oxide as an alternate terminal electron acceptor. Uninoculated controls remained passive with the anodic current decreasing steadily with exposure. The cathodic reaction was polarized during these experiments under both inoculated and uninoculated conditions (8) Iron oxidizing bacteria which oxidize ferrous iron to ferric iron sometimes promote the corrosion of iron and stainless steel pipes in aerobic environments. Corrosion products composed of ferric hydroxides and other metal salts form tubercles which accumulate on the inner surface of the pipes. The area beneath the deposit becomes anaerobic due to the oxygen diffusion barrier created by the precipitate and the respiratory activities of the bacteria. Bacteria that have been commonly found in tubercles over pits in stainless steel include *Sphaerotilus natans, Gallionella* and *Siderocapsa* spp..

Metal Reduction

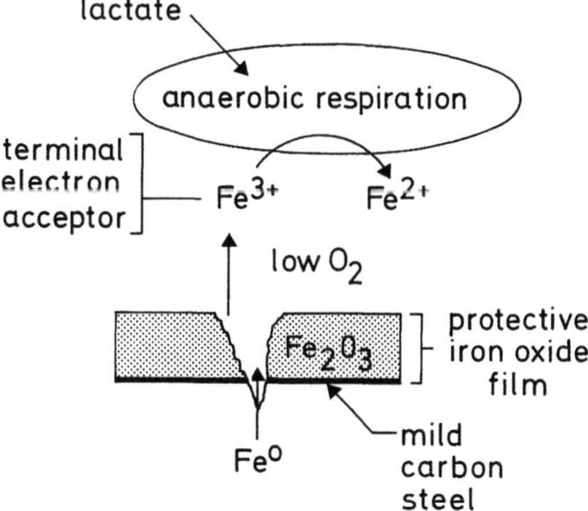

Fig. 8. Iron production by anaerobic respiration of ferric iron associated with protective gamma ferric oxide film on mild carbon steel

Bacteria such as *Gallionella* and *Sphaerotilus* act to produce or concentrate oxidizing species such as Fe^{3+} or Mn^{4+} at the metal surface (Fig. 9). On stainless steels in the presence of chloride ions this increases the potential to a value more noble than the pitting potential and leads to breakdown of a passive oxide film and promotes pit formation (Fig. 9).

Metal Oxidation + Precipitation

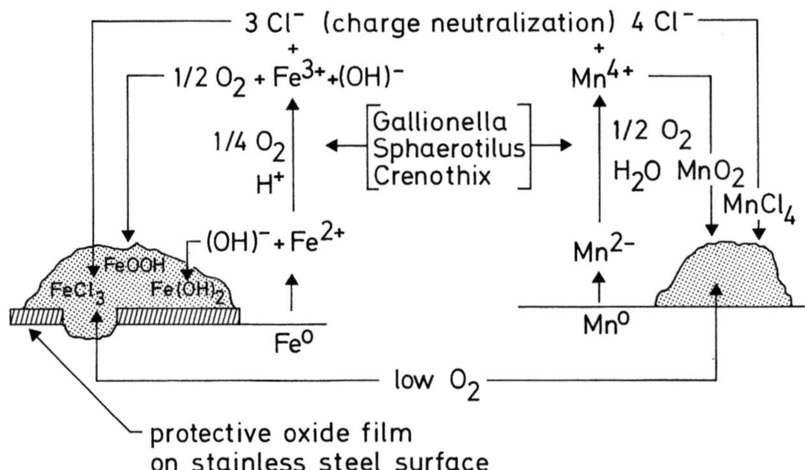

Fig. 9. Iron and manganese oxidation and precipitation in presence of filamentous bacteria. Stainless steel pitting in presence of chloride ions concentrated at surface in response to charge neutralization of ferric and manganic cations

Role of Biofilm Bacteria in the Establishment of Metal Concentration Cells

Anodic "depolarization" may occur as a result of the formation of metal ion diffusion gradients in biofilms. Reduced metal ion activity within localized regions of a biofilm can be mediated by bacterial processes that result in metal ion uptake or binding at the metal surface. Many species of bacteria, algae and fungi are known to bind various metal ions through both metabolically-energized and non-metabolic processes (9).

Role of exopolymers in the establishment of surface heterogeneity

Chemical variations among microcolonies of physiologically distinct microorganisms may occur as a result of differences in the chemical properties of the exopolymers excreted by the cells. Exopolymers are elaborated by most aquatic bacteria, cyanobacteria and algae. They form a matrix between the cells in a microcolony and in many instances anchor the cells to solid surfaces. The chemical properties of the exopolymers vary depending on the type of microorganism. Most water bacteria produce acidic exopolysaccharides which vary in their tendency to ionize and interact with metal ions (10). When one species of bacteria which excretes exopolymer with a high affinity for ions of the underlying metal forms a microcolony next to another bacterial species which produces exopolymer with low affinity for the metal ion, a metal concentration cell may develop; the area under the exopolymer with high affinity being anodic to that under the exopolymer with low affinity (11). Different interactions between exopolymers and copper ions have been proposed to lead to the formation of a copper concentration cell on a copper metal surface containing a mixed species microbial biofilm (12) The exopolymers also hinder diffusion of excreted products of microbial metabolism out of the biofilm. The concentrations of some particularly corrosive products such as those described above may increase to sufficiently high levels to promote deterioration of the underlying metal surface. Some types of pitting corrosion are believed to occur as a result of excretion and accumulation of microbial metabolites in localized regions of biofilms that have developed on copper surfaces (13)

CONCLUSIONS

At present, the importance of microorganisms in the corrosion of metals is not known. Circumstantial evidence provides of the bulk of evidence for a role for microbes in corrosion reactions. As our understanding of biofilm processes improves, it should become easier to relate specific biochemical activities on surfaces to electrochemical reactions that cause metal corrosion. This understanding will likely develop through a combination of highly-controlled studies in the laboratory and examination and characterization of corrosion events in the field.

REFERENCES

1   von Wolzogen Kuhr CAH, van der Vlugt IS (1934) The graphitization of cast iron as an electrochemical process in anaerobic soils. Water 18, 147-165

2   Little B, Wagner P, Gerchakov SM, Walch M, Mitchell R (1986) The involvement of a thermophilic bacterium in corrosion processes. Corrosion 42, 533-536

3   Costerton JW, Geesey GG (1986) The microbial ecology of surface colonization and of consequent corrosion. In: Dexter SC (ed) Biologically Induced Corrosion, NACE, Houston, TX; 223-232.

4   Wanklyn JN, Spruit CJP (1952) Influence of sulfate-reducing bacteria on the corrosion potential of iron. Nature 169, 928-929.

5   Walch M, Ford TE, Mitchell R (1989) Influence of hydrogen-producing bacteria on hydrogen uptake by steel. Corrosion 45, 705-709

164

6   Little B, Wagner P, Duquette D (1988) Microbially-induced increase in corrosion current-density of stainless steel under cathodic protection. Corrosion 44, 270-274

7   Obuekwe CO, Westlake DWS, Cook FD, Costerton JW (1981) Surface changes in mild steel coupons from the action of corrosion-causing bacteria. Appl. Environ. Microbiol. 41, 766-774

8   Westlake DWS, Semple KM, Obuekwe CO (1986) Corrosion by ferric iron-reducing bacteria isolated from oil production systems. In: Dexter SC (ed) Biologically Induced Corrosion, NACE, Houston TX; 193-200

9   Geesey GG, Jang L (1989) Interactions between metal ions and capsular polymers. In: Beveridge TJ, Doyle RJ (eds) Metal Ions and Bacteria, Wiley, NY; 325-257

10   Ford, TE, Maki JS, Mitchell R (1988) Involvement of bacterial exopolymers in biodeterioration of metals. In: Houghton DR, Smith RN, Eggins HOW (eds) Biodeterioration 7, Elsevier, New York; 378-384

11   Geesey GG, Jang L, Jolley JG, Hankins MR, Iwaoka T, Griffiths PR (1988) Binding of metal ions by extracellular polymers of biofilm bacteria. Wat. Sci. Tech. 20, 161-165

12   Geesey GG, Mittelman MW, Iwaoka T, Griffiths PR (1986) Role of bacterial exopolymers in the deterioration of metallic copper surfaces. Mat. Perform. 25, 37-40

13   Geesey GG, Bremer PJ (1990) Application of Fourier transform infrared spectrometry to studies of copper corrosion under bacterial biofilms. Mar. Tech. J. (in press)

CASE HISTORIES: BIOCORROSION

Robert E. Tatnall
DuPont Company
P.O. Box 80173
Wilmington, Delaware 19880-0173

ABSTRACT

Biocorrosion is a well-established, highly destructive phenomenon. Published cases link bacteria and fungi to accelerated corrosion of steel and cast iron, copper alloys, stainless steels, aluminium and nickel alloys. In addition, microorganisms can cause the destruction of plastics, stone, concrete and, of course, wood. While the exact figures are not available, biological factors probably play a role in about half of all the corrosion and materials degradation that occurs in the world.

This is a brief discussion of a variety of biocorrosion cases which occurred in industrial water systems. The focus will be on cast iron - where biocorrosion is commonplace and usually not properly diagnosed - and on stainless steels, where corrosion failures are, for the most part, unexpected and very expensive.

INTRODUCTION

Biocorrosion, or corrosion initiated or accelerated by microorganisms, is now recognized as a major destructive force in aqueous systems. First recognized by Dutch and British scientists in the late 1800's, biocorrosion is still, to this day, being observed yet misinterpreted by corrosion engineers and water treatment specialists. What many now understand as "tuberculation" corrosion of steel, caused by an active synergism between several different types of bacteria, is still simply called "water corrosion" or "under-deposit corrosion" by those who do not understand (or believe in) the biological factors.

The truth is, biocorrosion is not uniquely mysterious, nor is it too complicated to understand. It is simply the manifestation of common physical and biological reactions. As microbiologists and corrosion scientists learn to understand each others' world, the net results of these combined factors become understandable.

While we are not yet at that point of total understanding, we are nonetheless learning to recognize the symptoms of biocorrosion. It seems that no two cases are exactly alike, yet one finds many common factors among the various cases studied. This paper is a brief discussion of eight different cases which were diagnosed as biocorrosion. It is important when learning from these cases that they represent many different types of bacteria, and the chemical and biological factors are very different. Biocorrosion is not one phenomenon. Rather it is any of <u>many</u> forms of biological processes which affect chemical reactions at a metal surface.

H.-C. Flemming · G. G. Geesey (Eds.)
Biofouling and Biocorrosion in Industrial Water Systems
Proceedings of the International Workshop on
Industrial Biofouling and Biocorrosion, Stuttgart, Sept. 13-14,1990
© Springer-Verlag Berlin Heidelberg 1991

166

Case 1: Localized Corrosion of Buried Cast Iron

It seems fitting to begin with corrosion of an underground pipeline, as this is the earliest reported type of biocorrosion.

Figure 1 shows part of an excavated pipeline of ductile cast iron. Alongside is seen one of the steel rods which hold these pipe sections together near 90 degree elbows. (The elbow is not visible in this picture.) Much localized corrosion can be seen on the pipe, although this was not the case when it was first excavated. As is typical in such cases, the freshly exposed pipe surface appeared "like new," with original surface (sometimes even including paint markings) apparently intact. The corrosion seen in Fig. 1 was apparent only after vigorously scrubbing of the surface with a wire brush. (Sandblasting is another effective method.) The reason that corrosion was suspected in the first place was because of the severe corrosion of the adjacent reinforcing rod.

Fig. 1. A partially excavated ductile cast iron pipeline, showing localized pitting-type corrosion attributed to sulfate reducing bacteria in the soil. Also visible are two steel reinforcing rods, the one on the left showing severe SRB-related corrosion. On steel such as this, the corrosion is readily apparent upon excavation. The cast iron, on the other hand, appeared "like new" until the soft graphite matrix remaining in the pits was removed with a wire brush. (Figure courtesy of W.K. Link, DuPont Co.)

The corrosion of both the cast iron pipe and the steel rod is due to action of sulfate reducing bacteria (SRB) in the soil. These bacteria create a highly reducing local environment by producing free sulfide. This is corrosive to ferrous metals, as it is to copper alloys. Because the corrosive environment is reducing, the ferrous portions of the cast iron structure dissolve away, but the graphite "skeleton" remains intact. This is why the pipe surface at first appears unaffected. On the mistaken belief that the whole cast iron matrix dissolves, with subsequent precipitation of graphite, past observers named this type of corrosion "graphitization." The steel rod, on the other hand, contains no such graphite phase, so its dissolution is more readily apparent (ie. it does not undergo "graphitization" corrosion.)

Exactly why such localized corrosion by sulfate reducers occurs is not totally clear. These organisms are *always* present in soil. Apparently, conditions for their growth become just right in certain cases. These would typically include the absence of oxygen (perhaps brought on by activity of other oxygen-consuming bacteria in the area), needed organic carbon supply (probably also a by-product of other bacteria), and sulfate.

Numerous references are available regarding soil-related biocorrosion (1-3). It has been determined that soil conditions along pipeline trenches usually differ significantly from the surrounding soil, and are likely to favor bacterial activity (4). Cathodic protection is generally effective in preventing biocorrosion of buried pipelines, but protective coatings and tape wraps on the pipes may actually provide nutrition for bacteria — and be consumed in the process (5). Disbonded coatings and tape wraps are ideal sites for bacterial activity, and can prevent protective currents from reaching the pipe surface. Coal tar and epoxy coal tar enamels are the protective coatings of choice for buried structures (6).

Case 2: Pitting of Steel Under Tubercles

In aerated water service, localized mounds are often seen on metal surfaces, and these mounds are commonly called "tubercles." Such tubercles frequently cover pits, which will, in time, completely penetrate the metal substratum. Tubercles may range in size from a couple of millimeters to several centimeters in diameter and height. The underlying pits are usually smaller in diameter than the covering tubercles.

On non-ferrous metals such as copper alloys, tubercles are usually soft and slimy before they dry out, and observers are easily convinced that they are largely biological in origin and makeup.

When tubercles grow on steel surfaces, however, they usually contain large quantities of iron corrosion products. As a result, they are usually hard and gritty in consistency. It is not surprising, therefore, that such tubercles have traditionally been thought to be inorganic deposits rather than biological "structures." In fact, while they contain only about 10 to 15% organic material, these tubercles are a mixed consortium of bacteria, large amounts of exopolymers (slime), and even larger quantities of iron and other inorganics bound by the exopolymers. They are very much a "living" structure, with strategic influx of things needed for bacterial growth, and rejection (or complexing and sequestering) of things not needed by the microbial community.

Near tubercle surfaces are found aerobic slime-forming bacteria — most often Pseudomonads. Scattered throughout the mound will often be found filamentous or stalked iron oxidizing bacteria. (Why these are present is not well established, but it may be that they are simply opportunists, using this rich source of reduced iron from corrosion as an energy source.) At the base of the tubercle, where oxygen is depleted, anaerobes flourish — including sulfate reducing bacteria (SRB).

Fig. 2 shows a typical pit associated with SRB activity. When the covering tubercle is first lifted away, the pit is generally filled with a black sulfide corrosion product. (One way to quickly confirm that it's a sulfide is to put a drop of 5-15% HCl on the black material. A "rotten egg" odor of H_2S confirms the presence of sulfide.) When SRB are active under a tubercle on steel, metal penetration rates of 2-5 mm/yr are common.

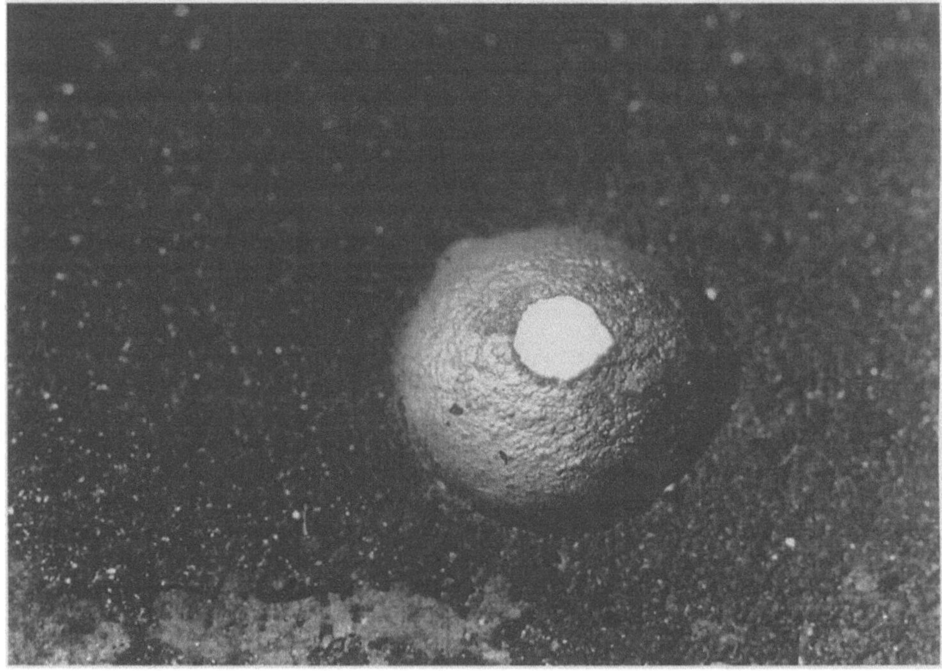

Fig. 2. A pit originating from the ID of a steel petroleum pipeline, and attributed to sulfate reducing bacteria. While not from the case cited herein, this pit is typical of pits in steel under biological mounds known as "tubercles." When the tubercle is first cleaned away, such a pit would initially be full of black corrosion products — primarily iron sulfide and phosphorus compounds. Actual pit diameter in this photo is about 1.5 cm. (Photo courtesy of G. Kobrin, DuPont Co.)

Proposed mechanisms for SRB-induced corrosion in aerated waters are discussed widely in the literature. Among those proposed are: Oxygen or redox differential cells generated under tubercles (with respect to surrounding surfaces), direct corrosiveness of hydrogen sulfide, cathodic effects of iron sulfide films in contact with the metal, protective properties of such sulfide films (leading to local corrosion activity when these films are breached), and cathodic depolarization by oxygen (remaining when the sulfur in sulfate is reduced) or by hydrogen-consuming enzyme activity associated with SRB and other bacteria (2, 3, 7-11). No definitive experiments have been reported which clearly state the relative importance of these factors. It is most likely that *all* of these factors are important to a greater or lesser extent in any given case.

Most discussions of SRB-related corrosion ignore a very important factor: The effects of *other* organisms working in concert with SRB. Only within the past few years have investigators started to report experiments with mixed bacterial consortia (12).

Case 3: Non-Pitting Biocorrosion of Steel Under a General "Scale"

Whereas the two previous cases involved very localized, pitting-type corrosion, not all biocorrosion fits this pattern. One example occurred in a small recirculating cooling water system in the southeastern U.S. This site had traditionally used chromates for corrosion control, but had been compelled for environmental reasons to switch to a zinc/polyphosphate corrosion inhibitor treatment. For several months thereafter, high corrosion rates were reported in carbon steel equipment. When a linear polarization corrosion probe was placed in the system, a hard "scale" formed on the steel electrodes within about a week (Fig. 3). When the crusty deposits were removed, a general, non-uniform corrosion pattern was seen, as shown in Figure 4. It was first assumed that this corrosion was due to an ineffective (or inadequate dose of) corrosion inhibitor, and biocorrosion was not even suspected.

The corrosion probes, as expected, indicated high corrosion rates on the order of 1.5 to 3 mm/year. More significantly, however, it was seen that the corrosion rate dropped very low twice a week, and then rose thereafter. It was also seen that doubling the corrosion inhibitor dose had no effect at all. The "low corrosion rate" incidents, it was determined, coincided with periodic additions of a non-oxidizing, quaternary amine biocide. This was the first clue that the corrosion was biological in origin. This was further confirmed when doubling of the biocide doses reduced corrosion by about half.

Further investigation revealed that the tower chlorinator had been taken out of service before the switch from chromates was made, and had never been put back on line because it was considered superfluous at that time. It seems the combination of chromates plus non-oxidizing biocide was adequate to control corrosion and fouling in the system. Chromate is a powerful biological control agent by itself. (Which is why it is so environmentally undesirable.) Obviously, the combination of *non*-chromate inhibitors plus minimal doses of quaternary amine biocide was *not* adequate to control biological activity. When the chlorinator was put back into service and doses of quaternary compound doubled, corrosion in the system dropped to less than .02 mm/year — and the "scaling" of surfaces stopped completely.

Fig. 3. The electrode end of a linear polarization corrosion probe after about one week's immersion in a cooling water system having inadequate biological control. The "scale" on these steel electrodes was hard and gritty, yet it was most likely biologically bound. In any case, the deposits no longer formed after a more aggressive biocide program was initiated.

This case illustrates how much can be accomplished by simply monitoring a problem system. Expensive analyses are not always required if a simple solution to a problem is all that is required.

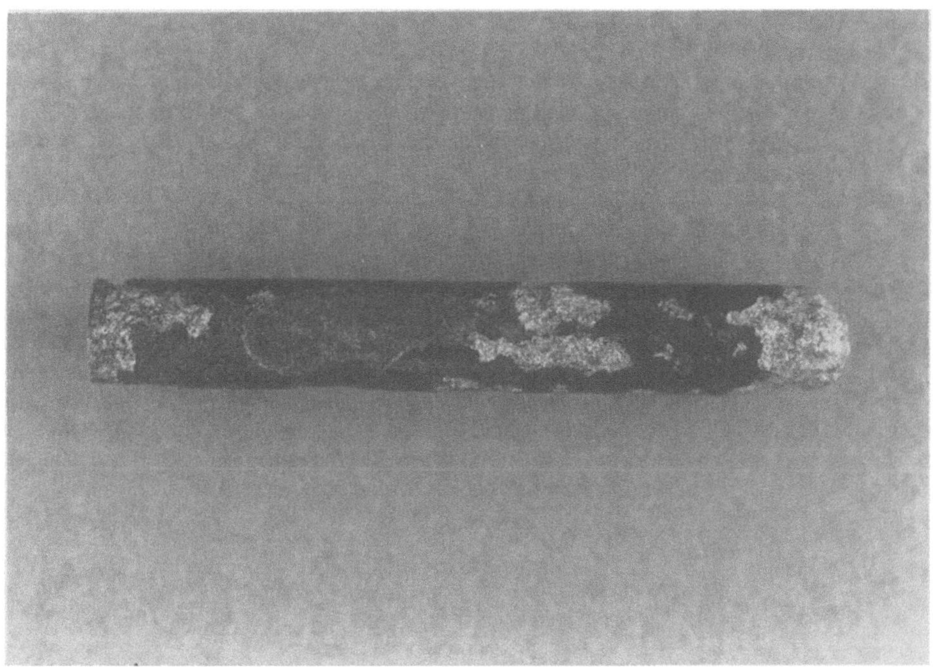

Fig. 4. One of the steel electrodes removed from the probe in Figure 3, shown after removal of the scale. Note that the corrosion is not pitting, but is somewhat localized in nature. This corrosion, as with the covering scale, stopped occurring after biological control of the system was improved. Earlier attempts to control it by increasing the dose of corrosion inhibitor proved futile

Case 4: Subsurface Pitting of Stainless Steels at Welds

The above cases involved steel and cast iron — metals which are <u>expected</u> to corrode at some rate anyway. Such metals and alloys rarely develop a truly protective coating on their surface (unless one is applied as a paint film), and so tend to oxidize continuously with time, the rate depending on moisture and other variables.

"Stainless" steels, on the other hand, get their name from a tendency to form a very protective oxide film, consisting primarily of iron and chromium oxides. This film forms spontaneously in the presence of air. As long as this film is intact, the alloy is said to be in a "passive" state, and will not corrode further at moderate temperatures. Oxidizing conditions (such as nitric acid or strong sulfuric acid) tend to reinforce this passive film, whereas reducing conditions (such as hydrochloric acid or, as we shall see below, hydrogen sulfide) tend to break it down, or "depassivate" the metal surface. Halogen ions are locally disruptive to this passive film, and stainless steels are generally prone to pitting corrosion in the presence of, for instance, high chloride ion concentrations. Perhaps the most destructive situation of all for stainless steels is high halogen ion concentrations in a strongly oxidizing medium. As will be seen here and again below, microorganisms are very capable of generating such a corrosive environment.

This case (13) involves a petrochemicals plant along the U.S. Gulf Coast, where a major process expansion was under construction. All piping and equipment for this project were stainless steel types 304 (18Cr-8Ni) and 316L (19Cr-9Ni-2Mo low carbon). This piping and equipment, once fabricated, was filled with well water and pressurized to test the integrity of welds and closures. After testing, the water was allowed to remain rather than being drained and dried out.

Within a few weeks, water began to leak from weld seams in the 304 piping, which was about 3 mm thick. Within four months, leaks had also penetrated similar pipe welds of type 316L. Inspection revealed internal reddish mounds along all of the weld seams, as shown in Fig. 5. These mounds were soft and slimy feeling when fresh and wet, but became hard and gritty when dried. They contained high numbers of *Gallionella* iron oxidizing bacteria and *Siderocapsa* iron and manganese concentrating bacteria. Elemental analysis of the mound material disclosed high levels of iron, manganese and chlorides.

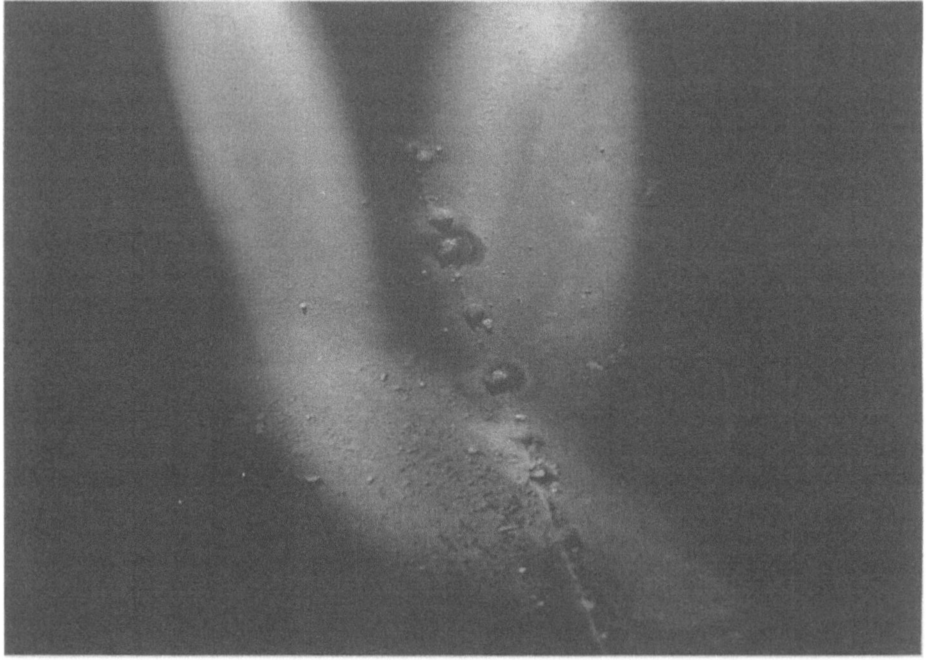

Fig. 5. Portion of a floor seam weld in a stainless steel storage tank. Visible are several biological mounds along the weld seam, each of which covered a subsurface pit. (Reprinted with permission: G. Kobrin, Mat. Perform. 15, p. 40.)

Fig. 6. Cross-section through a weld like that in Fig. 5, showing the subsurface nature of the pits in the weld seam. Note also that a smaller pit is visible some distance away from the weld. This is unusual in this particular case, and such pits probable coincided with weld spatter or other weld-related contamination. (Reprinted with permission: G. Kobrin, Mat. Perform. 15, p. 41.)

Under each of these mounds was found a subsurface pit, as shown in cross-section in Fig. 6. These pits were not visible when the mound was first wiped away, as the pinhole at the surface was very small. Heavy sandblasting was used to open up the pits so they could be weld-repaired. (Radiography has also been used successfully to inspect for such pits.). Failures of this sort have been reported at numerous sites world-wide. As a result, the National Association of Corrosion Engineers (NACE) and other societies are recommending that water used for testing stainless steel equipment be of the highest possible purity (distilled water or steam condensate if possible), and that it not be left in the piping or tanks any longer than necessary. Draining and wiping dry (or blowing dry with air) after 3 to 5 days is highly recommended. This is especially important where ambient temperatures are above about 25 °C, as all failures reported to date have occurred above that temperature. Just why this type of corrosion occurs almost exclusively along welds is not fully understood. It has been determined, however, that the structural or chemical inhomogeneity inherent in unannealed welds is an important contributing factor, and that full thermal anneal can greatly reduce the tendency for this to occur (14).

Case 5: Subsurface Pitting of Stainless Steel <u>Not</u> At Welds

Whereas the previous case involved pitting specifically at stainless steel welds, there have been numerous reports of a different but related phenomenon which occurs in condenser tubes but has no relationship to welds.

One such case occurred in four shell and tube condensers which had water flowing through the tubes. Tube material was type 304 stainless steel, welded but fully recrystallized (cold drawn and then solution annealed). Temperatures were moderate at the metal surface − typically 30 to 40 °C.

All surfaces exposed to water in the condensers showed an accumulated slime film, typically 0.3-1 mm thick (Fig. 7).

Fig. 7. General slime fouling on a stainless steel exchanger. This biofilm, though it appeared "typical" for fresh-water fouling, was found to contain a high level of manganese. The concentration of this element is attributed to (unidentified) bacteria in the film. When reacted with chlorine in the water, this manganese is blamed for widespread, tunneling pitting in the type 304 stainless steel tubes

This biofilm was a gray-brown color, without any unusual visual characteristics. Analysis showed it to be high in manganese, however — typically 2 to 6 percent by weight in the dried biofilm. This was surprising, as the water typically contains only about 30 to 60 ppb (μg/l) manganese. The film was also high in iron, but this is typical for fresh water biofilms. No distinct chloride peak was seen using EDS (EDXA) analysis, so chloride levels in the film are believed to have been less than about 0.1% on a dry weight basis. The water had been continuously chlorinated at a fairly high level (0.5 to 1 mg/l free residual).

Scattered throughout the tube bundle, underneath the biofilm, were found sub-surface, "tunneling" pits as typified by Fig. 8. The tube metallurgy was determined to be normal for properly heat treated stainless steel. The pit morphology itself offered no clues as to the chemistry involved, in that there was no selective attack of any portion of the metal grain structure.

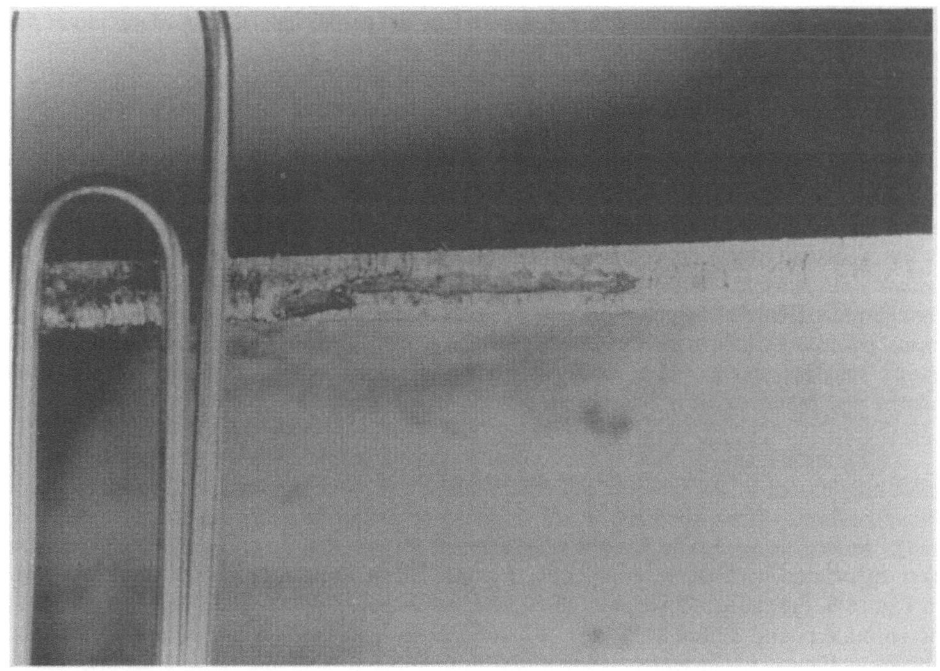

Fig. 8. Cross section of one of the tunneling pits in a stainless steel tube. While typical of the corrosion reported herein, this photo actually came from a similar case at another site. Tube wall thickness was 1.57 mm. The paper clip is included for size reference. (Photo courtesy of G. Kobrin, DuPont Co.)

The clue to this case turned out to be the manganese concentrated in the biofilm. New tubing identical to that effected here, when coated with a paste of MnO_2 and immersed into distilled water into which chlorine gas was slowly bubbled, developed identical subsurface tunneling within two weeks at room temperature. Another tube so exposed but without the manganese paste showed no pitting nor corrosion in the same time frame.

In the failed condensers, the microorganisms probably did nothing more than concentrate manganese from the water, and the free chlorine in the water did the rest. The corrodent was most likely permanganic chloride.

Several methods of preventing this problem come to mind. First, any biocide which will prevent the formation of the biofilm in the first place should thus prevent manganese concentration. (Obviously chlorine alone was <u>not</u> enough in this case.) Second, substituting a non-halogen biocide in place of chlorine should prevent formation of corrosive chemistry, even if the biofilm still forms. Thirdly, a system using continuous mechanical cleaning of the tube surfaces, either with reciprocating brushes or abrasive foam balls, should prevent the formation of a biofilm without the need for more expensive biocides. Both such systems are commercially available, and carry a much lower price tag than do repeated tube failures.

Case 6: Crevice Corrosion of Stainless Steel Under Gaskets

Type 304 stainless steel has been used increasingly for cooling water service in the chemical, petrochemical and power utility industries. This has often been done to eliminate the need for corrosion inhibitors (with their environmental concerns). In fact, it has spawned many new types of corrosion failures for which this common alloy is *uniquely* susceptible. What is worse, these problems often defy correction or remediation by chemical treatment — either by corrosion inhibitors or biocides. The two previous cases involved very corrosive chemistries which would corrode most alloys, including fairly exotic stainless steels. This case, on the other hand, might be solved by substituting almost any metal or alloy other than type 304.

Piping for this particular recirculating cooling water system was type 304, as were most condensers in the system. All flange gaskets were originally specified as elastomer--bound asbestos fiber, then the world standard for water and steam service. After a few years' service, several pipe flanges were opened for inspection, and crevice corrosion was seen associated with discrete biological mounds along the flange/gasket edge. As shown in Figure 9, the corroded site was filled with black corrosion products consisting primarily of chromium and sulfur (15, 16). As time passed, corrosion progressed to the stage shown in Figure 10, at which point it was impossible to seal the flanges against leakage.

A short-term solution to the problem was to clean out the corroded sites, fill with an epoxy putty, and reseal the flanges with soft elastomer (EPDM) gaskets. Later studies showed that the corrosion was caused by sulfate reducing bacteria, actively growing under mounds of aerobic slime-forming bacteria (16). As with tubercles on steel, discussed above, these deposits also contained some filamentous iron bacteria, although these are not considered important to the corrosion scenario.

This problem has defied solution by chemical treatment, including extremely high levels of oxidizing and non-oxidizing biocides. For example, when formaldehyde was added to a test system at a continuous dose rate of 5000 mg/l, the biological consortium simply changed — and new organisms apparently capable of utilizing formaldehyde appeared. Their metabolic byproducts included formic acid, and the corrosion merely got worse!

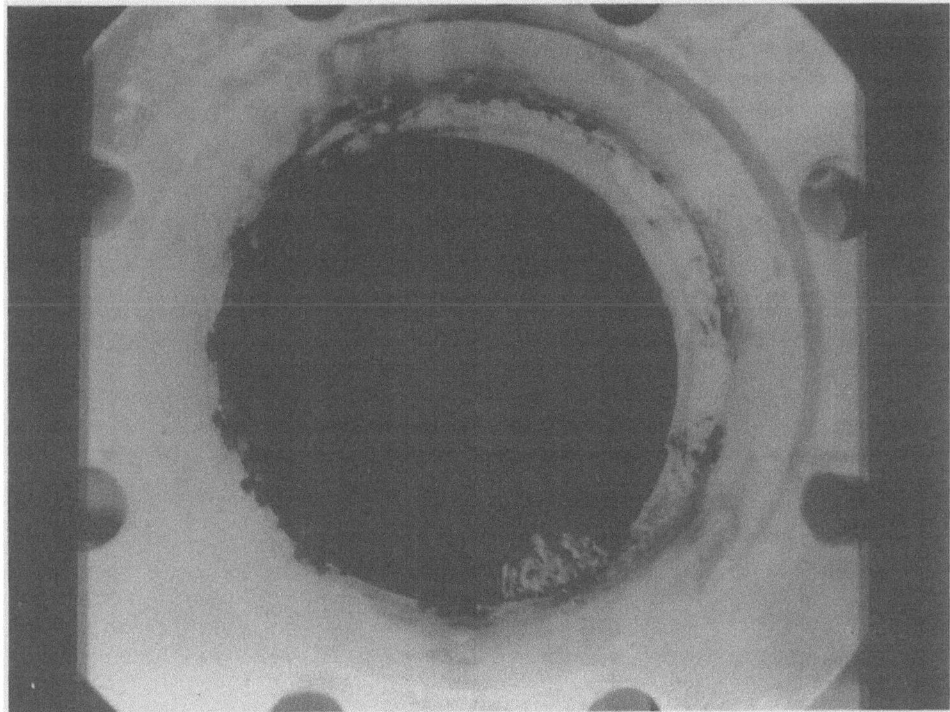

Fig. 9. A type 304 stainless steel pipe flange that had been mated to a blind flange in a recirculating cooling water system. Shown after about ten weeks' service. Note the soft, brownish biological mound near the top, which covered an area of crevice corrosion. The corroded site is filled with a black chromium/sulfur product. Each of the corrosion sites around the flange perimeter was covered with an identical mound when the system was first opened. (Reprinted with permission: R.E. Tatnall, Materials Performance, 20, 1981, p. 41.)

While "better" chemical treatment may be futile in this particular case, an easy long-term solution is to substitute type 316 stainless steel, which contains two percent molybdenum. While this slightly higher alloy is not totally immune to this corrosion, the rate of penetration is low enough that operating problems should never be a concern. Again, a distinction must be drawn between this SRB-related case, where the molybdenum addition is virtually a "cure," and the cases discussed earlier where oxidizing chlorides are the corrodent. In those cases, molybdenum addition only extends life by about a factor of two, so is not an ultimate cure. All of this illustrates the futility in attempting to define a list of "biocorrosion-resistant" alloys. It is essential that each case of biocorrosion be understood, and that alloys suitable for *that particular* situation be used.

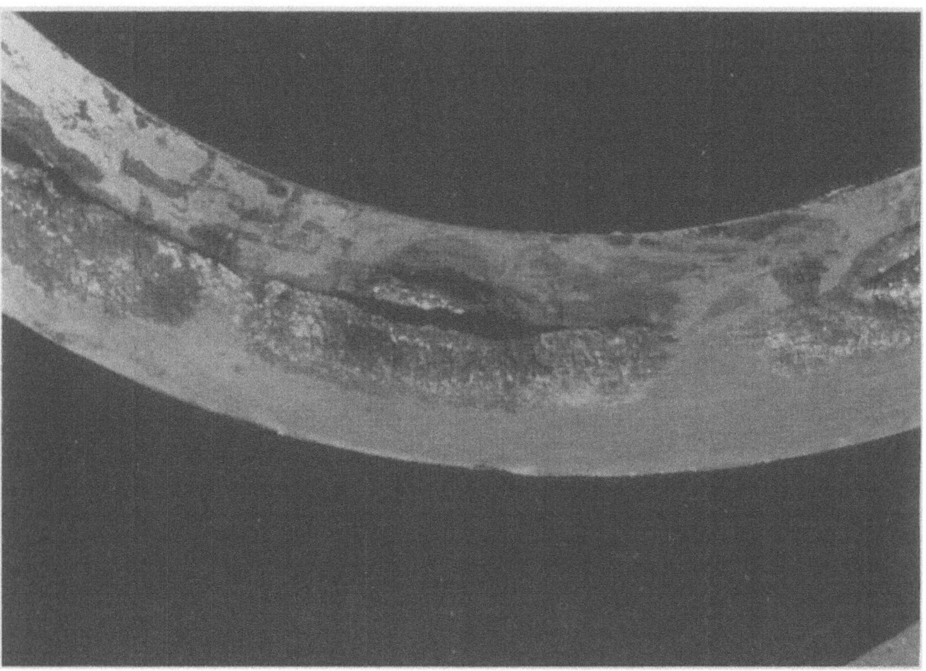

Fig. 10. More advanced crevice corrosion of the type shown in Figure 9. This corrosion measures about 1 cm in depth, and prevents sealing the flange against leaks. (Reprinted with permission: R.E. Tatall, "Experiments in Microbiologically Influenced Corrosion." Presented at NACE Corrosion '84, Paper No. 95, p. 14.)

Case 7: Simple Pitting of Stainless Steel Under Chloride-Rich Biodeposits

Many people who do not believe that microbes cause corrosion, refer to biocorrosion as simple "under-deposit" corrosion. When it's pitting of stainless steels that is involved, they look for the presence of chlorides, and then say the corrosion is simply due to "chloride concentration in deposits." The fact is, no one has proposed a mechanism for "simple under-deposit corrosion" which excludes a likely role for microorganisms. Similarly, no one has yet explained why chloride ions from the water should concentrate within a deposit unless local boiling is occurring — unlikely at the moderate temperatures usually associated with water side pitting corrosion. As shown in this case, the role of bacteria is not always well understood. When corrosion occurs *only* in their presence, however, it seems reasonable to conclude that the bacteria play some active role in initiating or propagating the corrosion.

Once again, this case involves a recirculating cooling water system and type 304 stainless steel. Here, the pitting corrosion occurs only under discrete biological mounds like that shown in Fig. 11. (Note that in this figure there is no general fouling of surfaces.) Under such mounds are found pinhole pits which, unlike earlier examples, penetrate straight through the metal, often as though they were produced with a drill.

Fig. 11. A type 304 stainless steel thermowell from a cooling water system, showing a discrete biological mound on an otherwise fairly clean surface. This soft, slimy mound was high in chloride concentration, which probably accounted for a straight, deep pit underneath. Why such biological communities concentrate chlorides is presently unknown

One of the more irregular examples of such pitting is shown in Fig. 12. There is nothing unusual about this pitting, except that it only occurs underneath these biological deposits. It otherwise appears like typical chloride pitting of stainless steel.

The mound material likewise shows nothing extraordinary. It typically contains high numbers of aerobic slime-forming bacteria, plus a measurable amount of chloride. In this case, there is no manganese. Also, "distinctive" morphological forms, such as iron oxidizing or sulfate-reducing bacteria, are generally not present in significant numbers.

Microbial ecologists generally agree that, within biological environments such as this, things don't "just happen," but almost always have a rational explanation that favors the survival of the organisms present. While this case involves more questions than answers, it appears that these bacteria, for some unconfirmed reason, gain an advantage by concentrating chlorides. Once again, however, since this does not involve the extremely corrosive ferric/manganic chloride situation discussed earlier, an upgrade to a molybdenum-containing grade of stainless steel should offer an effective cure. A more aggressive biocide treatment might also control the problem, but, as described earlier, thick, discrete mounds like these tend to resist biocide treatment better than does a thin biofilm.

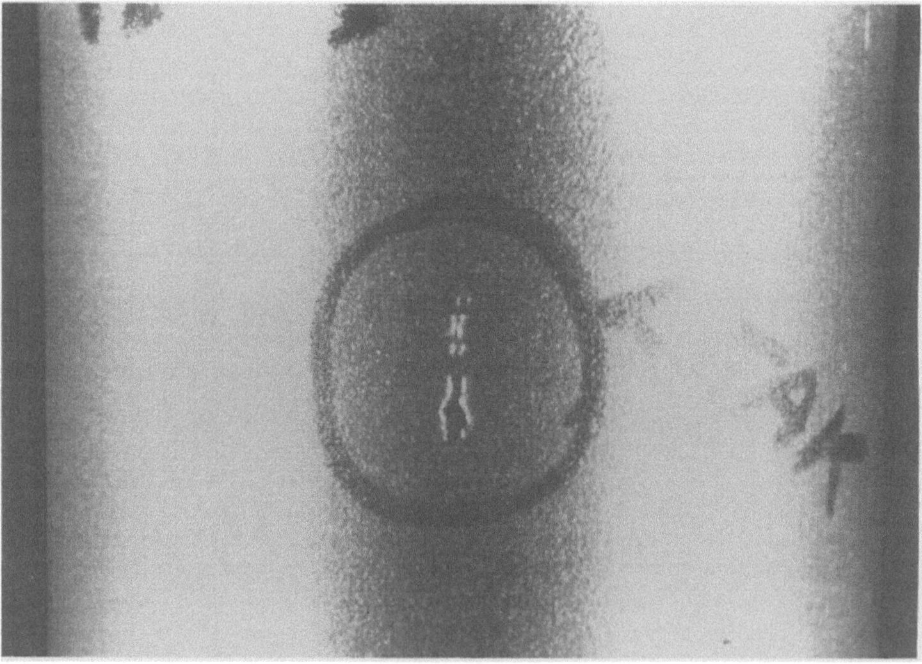

Fig. 12. A somewhat irregular version of a pit which occurred under a discrete mound like that shown in Fig. 11. Unlike the pitting described in other cases herein, these pits bore straight through the metal at a rate of about 0.5-1 mm/yr

Case 8: Biodeterioration of Concrete Sewer Pipes

This case is included to illustrate two points: 1) Not all "biocorrosion" involves metals, and 2) not all biocorrosion occurs in remote, industrial situations. This case, in fact, intimately involves all of us who live in urban areas of the civilized world. It also demonstrates one of the ironies of our times: An environmental problem *created* by dealing with another environmental problem!

Virtually all major sewer pipes are made of concrete. The largest of these, the gravity mains in major cities, measure up to about 5 meters in diameter. Most of these pipes are reinforced with steel for added strength. Gravity lines typically run about one fifth full of liquid and solid sewage, with the remaining four fifths being occupied by various gases, including air which enters through manways and other openings. Among the more noxious gases in sewers is hydrogen sulfide, produced by sulfate reducing bacteria in the slime and flowing liquid at the bottom of the lines. Historically, the sulfide concentration in the head gases has been on the order of 10 mg/l, and deterioration of the piping has not been a problem. Cities also used to include all sorts of waters in the sewers, including storm runoff. This kept flow rates high, and tended to flush the sewage out rather quickly.

More recently, municipalities have been forced to implement two operating changes in order to properly treat the sewage before discharging it to rivers or the ocean. First, they have separated out such things as storm runoff wherever possible − to minimize flow rates and the amount of material that must flow through their waste treatment system. This has greatly reduced average flow rates, increased residence time in the lines, and increased concentration of all reaction products including hydrogen sulfide. Secondly, industries are being forced to stop discharging toxic materials, including heavy metals, into the sewers. These things either interfere with the ultimate waste treatment, or pass right through it and pollute the receiving bodies of water. Those toxic materials, however, were also helping to limit biological activity in the sewers themselves. Systems which have been so "cleaned up" are experiencing tremendous increases in the sulfide level. In some cases, where there are sufficient sulfate levels in the stream to support high SRB activities, sulfide levels are in excess of 500 mg/l in the sewers − a level which will kill humans in minutes.

Also living in the sewers are sulfide oxidizing *Thiobacillus* bacteria. These organisms derive energy by oxidizing sulfide to sulfate. Because *Thiobacillus* require oxygen, they grow only in the upper portion of the pipes. Some strains produce sulfuric acid at a pH of 1 or less! Needless to say, this is quite destructive to both concrete and to reinforcing steel (17, 18).

As a result, some cities which have historically had no deterioration of sewers in more than one hundred years, have experienced complete destruction of parts of their systems within only the past few years. It is not unusual in such cases to see earth instead of concrete along the tops of the lines. Numerous collapses have occurred as a result. The city of Hamburg, Germany, reputedly lost a city bus in one of their cave-ins!

182

Answers to this problem are not easy and do not come cheap. Acid-resistant cements have been useful for spot repair, but taking lines out of service for this work is difficult — especially in the case of main trunk lines. Total replacement with plastic piping is a possibility, but the cost is usually prohibitive, as the mains are often buried ten meters or more beneath buildings, railroads and highways. A number of cities dump iron salts into their sewers to precipitate the sulfide out as iron sulfide — but this is expensive and creates voluminous added "sludge" which must eventually be buried in sanitary landfills.

Cases Involving Other Metals and Alloys:

Only cases involving iron, steel, stainless steel and concrete have been discussed here. There are documented cases of biocorrosion of most common materials of construction, however, including copper and copper alloys (19-22), nickel (20, 23), Monel (13, 24), aluminum (25, 26), "high alloy" stainless steels (27, 28), and in fact nearly all commonly used metals and alloys with the exception of titanium and Ni-Cr-Mo superalloys such as alloy C276. An overview of corrosion of these materials was published by the Materials Technology Institute of the Chemical Process Industries, Inc. (MTI) in 1984 (29).

CONCLUSIONS

These few cases are not by any means representative of the total diversity of biocorrosion of metals and biodeterioration of non-metals. Rather, it is a sampling to show the sorts of observations and analyses that are traditionally applied to suspected cases. It should also give the reader some ideas as to how to recognize biocorrosion symptoms.

Several general points have been illustrated by these few cases, including:

1. Biocorrosion often occurs in systems which are treated with biocides. In some cases, improved chemical treatment may solve the problem. In other cases, however, no quantity or quality of chemical treatment seems to help. In such cases, mechanical cleaning must be relied on to prevent the accumulation of biological materials on surfaces — or the materials of construction upgraded.

2. Biocorrosion very often occurs under discrete mounds or tubercles, and often in the complete absence of general, film-type biofouling. Sometimes, however, localized biocorrosion occurs under a general biofilm.

3. Biocorrosion takes many different forms, and involves many different chemistries and organisms. It is dangerous, therefore, to make general statements regarding conditions under which biocorrosion will or will not occur, or about which alloys are susceptible or resistant to biocorrosion.

One point perhaps not made clearly enough with these particular cases is that *biofouling* and *biocorrosion* are two very different, and often separable, things. It is not uncommon to have severe biofouling with no corrosion at all. Chances are, however, that biocorrosion will not occur in the absence of biofouling — even if the fouling film or deposit is too thin to be seen by the naked eye. Where corrosion occurs under a discrete mound or tubercle, it is most likely the conditions generated within that discrete mound which are causing corrosion, even if the discrete mounds are accompanied by a heavy biofilm.

REFERENCES

1   Fischer KP (1981) Cathodic Protection in Saline Mud Containing Sulfate Reducing Bacteria. Mat. Perform. 20, 41-46.

2   Miller JDA, Tiller AK (1970) Microbial Corrosion of Buried and Immersed Metal. In: Miller JDA (ed) Microbial Aspects of Metallurgy. Am. Elsevier Pub.; 61-105

3   von Wolzogen Kühn CAH (1937) The Unity of the Anaerobic and Aerobic Iron Corrosion Process in the Soil. Presented at the Fourth National Bureau of Standards Soil Corrosion Conference, Washington, D.C., 1937

4   Harris JO (1969) Soil Microorganisms in Relation to Cathodic Protection. Presented at NACE National Convention, Dallas, Texas, March 1960.

5   Lederer SJ (1966) Microbial Attack — The Scourge of Plastics and Coatings. Chem. Proc. (For Operating Management) March 1966; 54-60

6   Garrity KC et al. (1989) Corrosion Control Design Considerations for a New Well Water Line. Mat. Perform. 28, 25-29

7   Boivin J et al. (1990) The Influence of Enzyme Systems on MIC. NACE Corrosion 90, Paper No. 128, Las Vegas, Nevada, April, 1990

8   Edyvean RGJ (1990) The Effects of Microbiologically Generated Hydrogen Sulfide in Marine Corrosion. Marine Technol. Soc. J., 24, 5-9

9   Hamilton WA (1985) Sulphate-Reducing Bacteria and Anaerobic Corrosion. Ann. Rev. Microbiol. 39, 195-217

10   Tatnall RE (1981) Fundamentals of Bacteria Induced Corrosion. Mat. Perform. 20, 32-38.

11   Videla HA, et al. (1990) Bioelectrochemical Assessment of Biofilm Effects on MIC of Two Different Steels of Industrial Interest. NACE Corrosion 90, Paper No. 123, Las Vegas, Nevada, April 1990.

12   White DC, et al. (1990) Microbially Influenced Corrosion of Carbon Steels. NACE Corrosion 90, Paper No. 103, Las Vegas, Nevada, April 1990

13   Kobrin G (1976) Corrosion by Microbiological Organisms in Natural Waters. Mat. Perform. 15, 38-43

14   Borenstein SW, White, DC (1989) Influence of Welding Variables on Microbiologically Influenced Corrosion of Austenitic Stainless Steel Weldments. NACE Corrosion 89, Paper No. 183, New Orleans, Louisiana, April 1989

15   Tatnall RE (1981) Crevice Corrosion of Stainless Steel in Recirculating Cooling Water. Mat. Perform. 20, 41-48.

16   Tatnall RE (1984) Experiments in Microbiologically Influenced Corrosion. NACE Corrosion '84, Paper No. 95, New Orleans, Louisiana, April 1984.

17   Bock E, Sand W (1990) Microbially Influenced Corrosion on Concrete and Natural Sandstone. Presented at the International Congress on Microbially Influenced Corrosion, University of Tennessee, Knoxville, October 7-12

18   Mansfeld F, Shih H, Postyn A (1990) Corrosion Monitoring and Control in Concrete Sewer Pipes. NACE Corrosion '90, Paper No. 113, Las Vegas, Nevada, April 1990.

19   Alanis I et al. (1986) A Case of Localized Corrosion in Underground Brass Pipes. in Dexter SC (ed) Biologically Induced Corrosion NACE, Houston, TX

20   Little BJ et al. (1990) Microbiologically Influenced Corrosion in Copper and Nickel Seawater Piping Systems. Marine Technol. Soc. J. 24, No. 3, Sept. 1990; 10-17

21   Little BJ et al. (1988) The Impact of Sulfate-Reducing Bacteria on Welded Copper-Nickel Seawater Piping Systems. Mat. Perform. 27; 57-61

22   Videla HA, et al. (1989) Biofouling and Corrosion of Stainless Steel and 70/30 Copper-Nickel Samples After Several Weeks of Immersion in Seawater. NACE Corrosion '89, Paper No. 291, New Orleans, Louisiana, April 1989.

23   Brennenstuhl A et al. (1990) The Effects of Biofouling on the Corrosion of Nickel Heat Exchanger Alloys at Ontario Hydro. Presented at the International Congress on Microbiologically Influenced Corrosion, University of Tennessee, Knoxville, Oct. 1990

24   Gouda VK et al. (1990) Microbial-Induced Corrosion of Monel 400 in Seawater. NACE Corrosion '90, Paper No. 107, Las Vegas, Nevada, April 1990.

25   Elphick JJ (1970) Microbial Corrosion in Aircraft Fuel Systems. In: Miller DJA (ed) Microbial Aspects of Metallurgy. Am. Elsevier Pub.; 157-172

26   Schmitt CR (1986) Anomalous Microbiological Tuberculation and Aluminum Pitting Corrosion − Case Histories. In Dexter SC (ed) Biologically Induced Corrosion. NACE, Houston, Texas; 69-75

27   Scott PJB, Davies M (1989) Microbiologically Influenced Corrosion of Alloy 904L. Mat. Perform. 28, 57-60

28   Stein A, Merlino C (1990) Susceptibility of a 6% Molybdenum Stainless Steel to Microbiologically Influenced Corrosion. Presented at the International Congress on Microbially Influenced Corrosion, University of Tennessee, Knoxville, Oct. 1990.

29   Pope DH, et al. (1984) MTI Publication No. 13 − Microbiologically Influenced Corrosion: A State-of-the-Art Review. Columbus, Ohio: The Materials Technology Institute of the Chemical Process Industries, Inc.

SULPHATE-REDUCING BACTERIA AND THEIR ROLE IN BIOCORROSION

W. Allan Hamilton

Department of Molecular and Cell Biology
Marischal College University of Aberdeen
Aberdeen AB9 1AS
United Kingdom

ABSTRACT

Sulphate-reducing bacteria (SRB) are the microorganisms most widely implicated in cases of biocorrosion arising in a wide range of natural and industrial environments. Models for their mechanism of action have concentrated on cathodic stimulation of the electrochemical process by hydrogen oxidation and/or the production of iron sulphide corrosion products. Preventatitive measures are largely confined to cathodic protection by sacrificial anode or impressed current, or the use of biocides in contained systems. Although SRB are strictly anaerobic organisms, they can be responsible for extensive biocorrosion under aerobic environmental conditions.

Over recent years two features characteristic of biocorrosion have become dominant in our effort to first understand and then control the processes involved.

a)   The physical and chemical nature of the iron sulphide corrosion products and in particular their interaction with oxygen, appear to determine the rate and extent of corrosion.

b)   SRB exist as components of complex microbial communities within a biofilm adherent to the metal surface. This biofilm is a dynamic structure composed of cells, extracellular polymeric substances (EPS), and inorganic inclusions including corrosion products. Many biological and chemical processes become diffusion-limited, and within the biofilm the presence of microenvironments is of great significance to both microbial activities and the electrochemical reactions of corrosion.

INTRODUCTION

Corrosion is a generic term covering a number of electrochemical processes (e.g. pitting, generalized weight loss, graphitisation, stress corrosion cracking, hydrogen embrittlement) which may affect iron, mild steel, stainless steels, and various aluminium, copper, nickel and cobalt alloys. Biocorrosion can involve a plethora of organisms and mechanisms; effects may be specific as with Fe^{2+} to Fe^{3+} oxidation by *Gallionella* or the production of organic acids by *Cladosporium*, or of a more general character as with differential aeration cells arising from colonial growth or biofilm patchiness (1). Although the nature of the corrosion process can generally be defined within narrow limits in respect of its

H.-C. Flemming · G. G. Geesey (Eds.)
Biofouling and Biocorrosion in Industrial Water Systems
Proceedings of the International Workshop on
Industrial Biofouling and Biocorrosion, Stuttgart, Sept. 13-14,1990
© Springer-Verlag Berlin Heidelberg 1991

metallurgy, the microbiological component is seldom simple or easily identified with a single organism or mechanism. This last fact is fundamental to our understanding of biocorrosion, but is only now beginning to be widely appreciated.

SULPHATE-REDUCING BACTERIA AND BIOCORROSION

Most widespread, distinctive and economically important is the biocorrosion resulting from the activities of the sulphate-reducing bacteria (SRB), which generally cause a pitting corrosion of mild steel characterized by the presence of black iron sulphide corrosion products (2).

Physiology and Nutrition

SRB are obligate anaerobes that use sulphate as the terminal electron acceptor for their respiratory metabolism, thereby producing significant amounts of sulphide. They have a limited nutritional spectrum, being generally unable to grow on carbohydrate sources or biopolymers, but preferentially utilizing short chain organic acids or alcohols; many species can additionally oxidize hydrogen as a source of metabolic energy and reducing power (3). These properties underlie both the environmental distribution of SRB, and their role in biocorrosion.

Generalized Mechanisms of Corrosion

Metal loss during corrosion occurs by the passage into solution of positively charged metal ions at the anode of an electrochemical cell. This must be balanced by transfer of the excess electrons to that part of the metal substratum which is cathodic and where a second reaction can absorb the electrons. The simple, non-stoichiometric, equations describing aerobic corrosion or rusting of iron would be:

$$Fe \rightharpoonup Fe^{2+} + 2e \qquad \textbf{anode}$$
$$1/2O_2 + H_2O + 2e \rightharpoonup 2OH^- \qquad \textbf{cathode}$$

with the subsequent formation of a complex of iron oxides and hydroxides as corrosion products.

In the absence of oxygen, it has been hypothesized that protons (4) or hydrogen sulphide (5) might serve as electron acceptor at the cathode:-

$$\text{a)} \quad 2H^+ + 2e \rightharpoonup 2H \rightharpoonup H_2$$
$$\text{b)} \quad 2H_2S + 2e \rightharpoonup 2HS^- + H_2$$

with SRB subsequently oxidizing the molecular hydrogen, thus both preventing polarization of the cathode by adsorbed hydrogen, and giving rise to sulphide and the potential for metal sulphide corrosion products:

$$4H_2 + SO_4^{2-} \rightarrow 4H_2O + S^{2-}$$
$$Fe^{2+} + S^{2-} \rightarrow FeS$$

That SRB can in fact oxidize cathodic hydrogen and use it as a source of metabolic energy has now been shown by several groups (6, 7, 8). In a complementary demonstration, it has been noted that although the application of impressed current can indeed confer protection against corrosion weight loss, it also stimulates SRB growth and activity in direct response to the increased tendency to produce hydrogen at the cathode (9).

Most discussions of biocorrosion by SRB however, have concluded that the major factor determining the rate and extent of corrosion is the presence of the iron sulphide corrosion products themselves (10). In purely electrochemical terms, iron sulphide is cathodic to unreacted iron, and the deposition of corrosion product might also be expected to offer an increased surface area for the formation and subsequent oxidation of cathodic hydrogen. More detailed studies however, have identified that thin adherent layers of the primary corrosion product mackinawite are generally protective, but that time-dependent rupture or high iron concentration-dependent bulky precipitation can lead to high non-transient corrosion rates which are independent of microbial growth or hydrogenase activity (11, 12).

Aerobic/Anaerobic Interface

Although SRB are strict anaerobes, it has been noted that in sediment systems they demonstrate maximal activity within a few centimetres of the aerobic/anaerobic interface (13, 14). Almost certainly this reflects nutrient input from the metabolic activities of other aerobic and facultative organisms also present but with greater biodegradative capability than the SRB, and the availability of biotic and abiotic oxidation of the produced sulphide, to which the SRB are themselves sensitive. Equally, extensive field observations have clearly identified increased risk of SRB biocorrosion in environments subject to intermittent access of oxygen. This last may be a direct effect on the sulphide corrosion products rather than resulting from any stimulation of SRB activity, as has been convincingly shown in both laboratory experiments (15) and in SRB-free geothermal power stations (16).

ENVIRONMENTAL STUDIES

A major limitation however, of the studies which have led to our present level of understanding of the mechanisms of SRB biocorrosion is that, for the most part, they have been carried out with pure strains of SRB in homogeneous monoculture with lactate as carbon and energy source, under controlled conditions of pH and Eh, and with the time course of most experiments being measured in days or weeks. What we have attempted to do in Aberdeen therefore, is to develop experimental systems that more accurately reflect those environmental conditions generally associated with corrosion in the field; this we have done with particular reference to the offshore oil industry.

Firstly we have been conscious of the fact that SRB within biofilms are component organisms within a nutritionally integrated microbial consortium and that consequently laboratory pure culture identification and enumeration is likely to give a very inexact picture of in situ activities. We have adapted a ^{35}S-sulphate reduction assay to enable offshore analysis of total SRB activity within undisturbed biofilms while still adherent to the metal surface of suitably sited and exposed test coupons (13, 17). Simple chemical analysis of sulphides on these coupons gives additionally a measure of total activity throughout the exposure period. Scanning electron microscopy, X-ray diffraction and energy dispersive X-ray analysis (EDXA) can give data on pitting characteristics and on the physical and chemical nature of the corrosion products. Such coupons were exposed, with and without cathodic protection by sacrificial anodes, for periods of from several weeks to two years, either attached to oil production platforms or sited on the sea bed in association with drill cutting sediments, or placed in laboratory simulations.

In summary, these studies have demonstrated the absence of any direct correlation between the extent of corrosion weight loss, with or without cathodic protection, and three measures of biological activity; SRB numbers, in situ rates of ^{35}S-sulphate reduction, or total sulphide corrosion products formed during the exposure period. Detailed analysis of the corrosion products was incomplete but in these longer term exposure studies we have found an association between thin adherent layers of iron sulphide, which often include pyrite, and low levels of corrosion; while loose, bulky films of iron oxides and sulphides, often including mackinawite, were associated with high, non-uniform corrosion rates. Furthermore, there seems to be some evidence that cathodic protection by sacrificial anode promotes the formation of thin adherent films of pyrite at the expense of bulky, loose deposits of mackinawite. Corrosion will then only occur if there is a breakup or discontinuity in this pyrite film. Since pyrite film is less prone to breakup than mackinawite, which is inherently porous and cracked, it is less likely to lead to corrosion.

The highest rates of corrosion (250-425μmy^{-1}, with characteristic SRB pitting) were associated with bulk aerobic environments where the thin adherent and bulky sulphide films were surmounted by a third layer of mixed calcareous deposits, and iron oxides and hydroxides. The lowest rate of corrosion (20μmy^{-1}) was found with samples buried in anaerobic sediment and characterized by a single adherent film of iron sulphide.

Corrosion Products and Oxygen

These studies therefore confirm that whereas SRB are the primary cause of the formation of sulphide corrosion products, the extent of the resulting corrosion is determined by the physical and chemical nature of these products. These, in turn, are critically modified by time- and environment-dependent factors, notably oxygen, in a manner that remains to be fully elucidated by further study specifically directed to this end.

BIOFILMS

A second major issue which must be more fully appreciated and understood, and which should be central to all considerations of methodology and mechanistic models is the fact that biocorrosion most usually results from the presence of microorganisms in the form of a biofilm at the metal surface. Biofilms, and their characteristic features, considerably influence the course of corrosion processes through both effects on microbial activities and, more directly, on the electrochemical reactions themselves.

The generalized biofilm will consist of a number of microbial species, either functionally independent or in some form of positive nutritional interdependence. The main bulk of the biofilm will however, be made up by the mixture of extracellular polymeric substances (EPS) characteristic of the incorporated species. Mass transfer of material into the biofilm, the metabolic activities of the constituent organisms, and the adsorptive and ion exchange properties of the EPS determine that diffusional processes dominate and that the biofilm as a whole is characterized by concentration gradients, giving rise to microenvironments and functional heterogeneities (18). Such heterogeneities can be increased in the horizontal dimension where patchiness arises from sloughing of mature biofilm or due to more colonial types of development.

These biofilm properties therefore, offer a ready explanation for the presence of SRB at the metal/biofilm interface, where the bulk medium may be aerobic and the principal nutrient source petroleum products. The presence within the biofilm of aerobic and facultative species capable of hydrocarbon degradation and subsequent fermentative metabolism will both generate the necessary oxygen-depletion within the deeper layers of the biofilm and supply suitable carbon and energy nutrients to support SRB sulphidogenesis and consequent biocorrosion. Equally, patchy or colonial growth can stimulate corrosion by the creation of oxygen concentration or differential aeration cells, the anaerobic area under the biofilm being anodic and the site of metal loss. This basic mechanism can then become further complicated by the presence of iron oxidizing bacteria stimulating the corrosion process by formation of insoluble ferric salts, and/or the development of SRB in the anaerobic regions beneath the other microbial growth.

Perhaps more subtle, but equally telling, are the direct influences of biofilm features on the electrochemical reactions of corrosion, and the experimental techniques used to monitor them. The classical electrochemical concept is of a metal/aqueous layer interface, with the possible additional complication of an inorganic passive film. What our knowledge of biofilms tells us is that to this model must be added consideration of the organic biofilm with its diffusional resistance leading to localised modifications in the type and concentration of ions, pH and redox values (19). With the onset of corrosion, the biofilm itself then becomes modified with the inclusion through entrapment of the inorganic corrosion products.

CONCLUSION

Biocorrosion arises from the reactions occurring within multi-component systems which are in dynamic flux. Both our conceptual models and experimental methodology must reflect and respond to these circumstances.

REFERENCES

1   Hamilton WA (1985) Sulphate-reducing bacteria and anaerobic corrosion Ann. Rev. Microbiol. 39, 159-217

2   Tiller AK (1982) Aspects of microbial corrosion In: Parkins RN (ed) Corrosion processes. Applied Science Publishers, London; 115-159

3   Widdel F (1988) Microbiology and ecology of sulfate- and sulfur-reducing bacteria In: Zehnder AJB (ed) Biology of anaerobic microorganisms. John Wiley, London; 469-585

4   von Wolzogen Kühr CAM, van der Vlught IS (1934) The graphitization of cast iron as an electrobiochemical process in anaerobic soils. Water 18, 147-165

5   Costello JA (1974) Cathodic depolarisation by sulphate-reducing bacteria. S. Afr. J. Sci. 70, 202-204

6   Pankhania IP, Mossavi AN, Hamilton WA (1986) Utilization of cathodic hydrogen by *Desulfovibrio vulgaris* (Hildenborough). J. Gen. Microbiol. 132, 3357-3365

7   Hardy JA (1983) Utilization of cathodic hydrogen by sulphate-reducing bacteria. Br. Corros. J. 18, 190-193

8   Cord-Ruwisch R, Kleinitz W, Widdel F (1987) Sulfate-reducing bacteria and their activities in oil production. J. Pet. Tech. Jan, 97-106

9   Guezennec J, Therene M (1988) A study of the influence of cathodic protection on the growth of SRB and corrosion in marine sediments by electrochemical techniques In, Sequeira CAC, Tiller AK (eds) Microbial corrosion 1. Elsevier Applied Science, London; 256-265

10 King RA, Miller JDA (1981) Corrosion by the sulphate-reducing bacteria. Nature 233, 491-492

11 King RA, Miller JDA, Wakerley DS (1973) Corrosion of mild steel in cultures of sulphate-reducing bacteria: Effect of changing the soluble iron concentrations during growth. Br. Corros. J. 8, 89-93

12   Mara DD, Williams DJA (1972) The mechanism of sulphide corrosion by sulphate-reducing bacteria In: Walters A M, Hueck-van der Plas E H (eds) Biodeterioration of materials. Vol 2. Applied Science Publishers, London; 103-113

13   Rosser, HR, Hamilton WA (1983) Simple assay for accurate determination of [^{35}S] sulfate reduction activity. Appl. Environ. Microbiol. 45, 1956-1959

14   Jørgensen BB (1988) Ecology of the sulphur cycle: oxidation pathways in sediments. Symp. Soc. Gen. Microbiol. 42, 31-63

15   Hardy JA, Bown J (1984) The corrosion of mild steel by biogenic sulfide films exposed to air. Corrosion 40, 650-654

16   Braithwaite WR, Lichti KA (1980) Surface corrosion of metals in geothermal fluids at Broadlands, New Zealand. In: Casper LA, Pinchback TR (eds) Geothermal scaling and corrosion. ASTM STP 717. Amer Soc Test Mat; 81-112

17   Maxwell S, Hamilton WA (1986) Modified assay for determining the sulfate reduction activity at metal surfaces. J. Microb. Methods 5, 83-91

18   Wilderer PA, Characklis WG (1989) Structure and function of biofilms. In: Characklis WG, Wilderer PA (eds) Structure and function of biofilms. John Wiley, Chichester; 5-17

19   Videla HA (1989) Metal dissolution/redox in biofilms. In: Characklis WG, Wilderer PA (eds) Structure and function of biofilms. John Wiley, Chichester; 301-320

BIOFILMS AND CORROSION

J. William Costerton and J. Boivin
Department of Biological Sciences
University of Calgary, 2500 University Drive N.W.
Calgary, Alberta T2N 1N4, Canada

ABSTRACT

Bacteria grow preferentially in biofilms. This mode of growth allows these organisms to set up highly structured, physiologically cooperative communities because they remain in stable juxtaposition to the colonized surface and to each other. Planctonic bacteria cannot establish these highly organized consortia. As a consequence, the focused bacterial biodeterioration of insoluble substrates and microbially influenced corrosion of metals are both dependant on biofilm formation. Furthermore, the biodeterioration of complex soluble organic substrates requires more than one species of bacteria; biofilm formation is therefore also a sine qua non in these processes. We must understand biofilms if we are to understand and control biodeterioration.

INTRODUCTION

Since the birth of microbiology in the times of Pasteur and Koch, the study of bacteria has traditionally been conducted on in vitro pure cultures rather than on the organism in contact with its habitat. Two centuries of intensive study have produced an exhaustive and very useful perception of the amazing genetic and physiological capabilities of bacterial cells; this perception has resulted in the production of a myriad of useful products such as vaccines, antibiotics, and genetically engineered organisms. In the past two decades, methods have been developed to allow the detailed chemical examination of bacteria in situ in their native habitats and we are led to the inescapable conclusion that phenotypic plasticity is actually the most amazing characteristic of the bacteria cell (1). It is now apparent that cells grown in pure <u>in vitro</u> cultures bear very little resemblance to cells of the same species growing in a specific natural habitat. Bacterial cells adapt themselves chameleon-like to virtually every microniche that they occupy. At almost the same time that M.R.W. Brown and his colleagues discovered phenotypic plasticity in bacteria, we began to use morphological methods to examine bacteria directly as they grew in a variety of natural systems.

DIRECT EXAMINATIONS OF BACTERIA

Early insights were developed by Marshall and his colleagues (2) who, in their examinations of the growth of marine bacteria, noted the pronounced tendency of these organisms to adhere to surfaces and then to divide to form adherent microcolonies and biofilms (Fig. 1)(3). When a wide variety of nutrient-sufficient natural and industrial

H.-C. Flemming · G. G. Geesey (Eds.)
Biofouling and Biocorrosion in Industrial Water Systems
Proceedings of the International Workshop on
Industrial Biofouling and Biocorrosion, Stuttgart, Sept. 13-14,1990

ecosystems had been examined quantitatively, it became clear that the majority of the bacterial cells in these systems actually grow in exopolysaccharide-enclosed adherent biofilms on available surfaces. An accurate estimate of the physiological activity of bacteria within an aquatic system could, therefore, only be obtained by including the direct study of intact mixed-species biofilms (Fig. 1). Meanwhile, direct examination of microorganisms in the very nutrient-poor zones of the deep oceans revealed yet another profound bacterial adaptation to oligotrophic conditions - the formation of very small (<0,3 μm) starved dormant ultramicrobacteria (4). There is very little resemblance between the full-sized planctonic (floating) cells studied conventionally in vitro pure cultures and the same organism grown in nutrient-sufficient or in nutrient-poor natural environments.

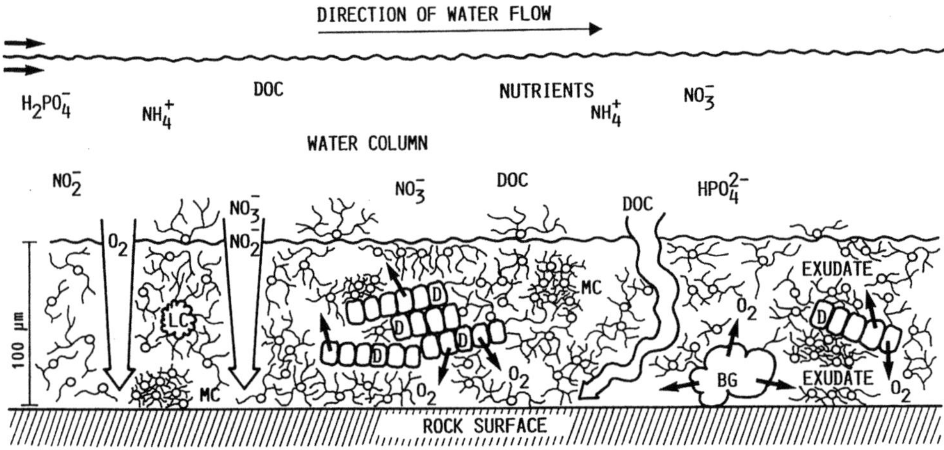

Fig. 1. Diagrammatic representation of a natural adherent biofilm in which bacteria (open circles) live within a continuous matrix of exopolysaccharide made by themselves and by their algal symbionts. The diagrams speculates concerning processes within this microbial biofilm, where diatoms and blue-green algae (cyanobacteria) are physiologically integrated with the adherent bacteria. *BG*: blue-green bacteria; *D*: diatoms; *DOC*: dissolved organic carbon; *LC*: lysed cyanobacteria; *MC*: microcolony

BIOFILM BACTERIA IN BIODETERIORATION

The conditions encountered by bacterial cells within a biofilm microniche differ radically from those in the bulk fluid of any given ecosystem. Unlike planctonic cells, they live within hydrated matrices where they remain in a restricted orientation to the colonized surface, to other bacteria in multispecies biofilms, and in close proximity to their own enzymes and exopolysaccharide products. Unlike higher eukariotic cells which are restricted to growth alone when not in their sexual reproductive phase, the essentially asexual bacteria within biofilms can survive, grow, and multiply continuously until a burgeoning microcolony occupies the favorable microniche.

Bacteria bring all of the advantages of plasticity and the biofilm mode of growth to bear when they initiate the complex processes of biodeterioration. Most of the bacteria in natural ecosystems adhere avidly to inert surfaces where they form physiologically active adherent biofilms (Fig. 1) whose cellular and matrix components trap organic and inorganic nutrients from the flowing water phase (Fig. 2). The efficacy of this trapping of nutrients by sessile bacterial biofilms is attested to by the efficiency of the trickling filter. Nutrient molecules are trapped by the biofilm matrix and made available by dissociation to the cells within the biofilm. Antibacterial molecules (surfactants, antibiotics) are also impeded and trapped by the same predominantly anionic fibers, protecting the biofilm cells (Fig. 2). The glycocalyx-enclosed mode of growth shields biofilm bacteria from antibodies, bacteriocins, bacteriophage and phagocytic cells such as amoebae and white blood cells (5).

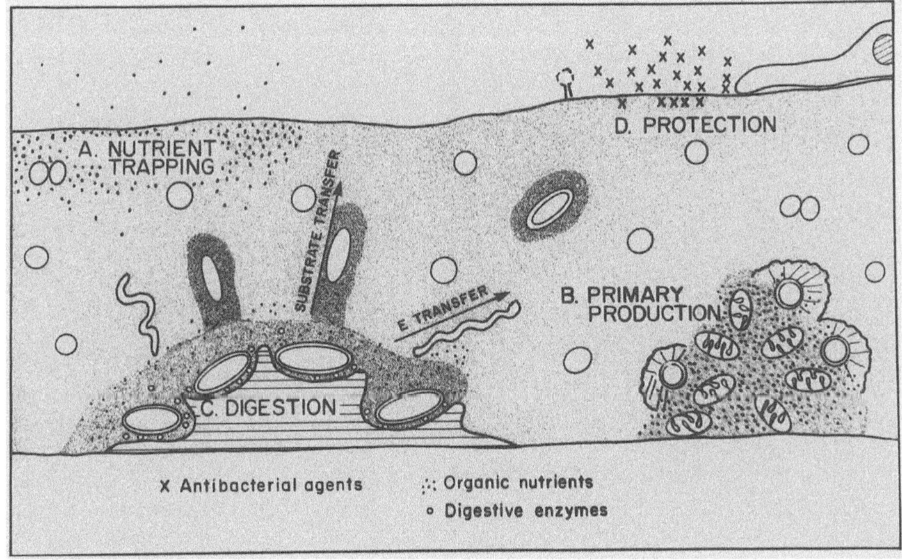

Fig. 2. Modes of nutrient acquisition and protection of a bacterial biofilm. *(A)* Nutrient trapping. *(B)* Primary production. *(C)* Formation of a digestive consortium. *(D)* Exclusion of antibacterial substances

The adhesion of bacteria with the capability of digesting specific materials often accelerates biodeterioration of these components. Cellulosic plants·materials are heavily colonized by cellulolytic bacteria (Fig. 3) within the first 15 minutes that they are in contact with rumen fluid (6). The cellulases and hemicellulases produced by these adherent bacteria are not released into the bulk fluid and their association with the exopolysaccharides of the developing digestive biofilm assures their eventual alignment at the cellulose to be digested (Fig. 3). Thus, both the degradative bacteria and their functional enzymes are retained at the surface that is undergoing biogradation and even fragments of the cell walls of dead bacteria can act as enzyme "platforms" and take part in cellulose digestion (Fig. 3). The effective degradation of solid materials could not be achieved by planctonic bacteria, because of the impossibility of their producing sufficient enzyme to saturate the bulk fluid. Biofilm focuses the attack on the surface to be degraded. Often the initial attack in a degradable surface involves the formation of a single species biofilm; this population will eventually be limited by the rate of metabolite removal. Cheng and his colleagues (7) have shown that cells of *Fibrobacter succinogenes* adhere avidly to cellulose and digest this substrate at a slow rate. However, when cells of a non-cellulolytic species of Treponema are added to the biofilm community, the metabolic end product of *F. succinogenes* cellulose digestion (butyrate) is removed and the rate of digestion is sharply accelerated.

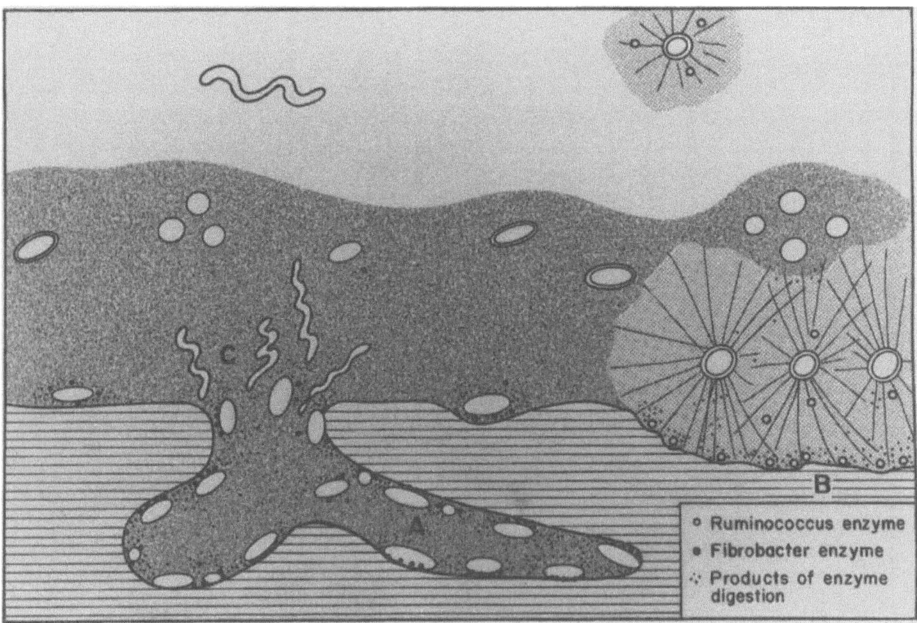

Fig. 3. In the bacterial digestion of cellulose, Ruminococcus cells remain at some distance from the substrate surface while their enzymes initiate digestion *(B)*. Other cellulolytic bacteria (e.g. *Bacteroides*) adhere intimately to the substrate *(A)* and detach vesicles and enzymes that cause a focal pitting attack. This attack is accelerated by the metabolic activity of noncellulolytic bacterial members of structured consortia

We predict that many natural biodeterioration systems will be shown to depend on the development of effective multispecies consortia, with their very efficient integrated physiological processes. These complex consortia cannot evolve unless the component organisms are immobilized and juxtaposed to each other within adherent biofilms. When the substrate of biodeterioration is insoluble (e.g. cellulose, wool) the degradation consortia will develop on its surface.

Even the degradation of soluble substrates often depends on bacterial consortia. The large microbiological particles that develop in wastewater treatment operation, notably in Upflow Anaerobic Sludge Blanket (UASB) reactors, are also composed of very complex microbial consortia (8). These consortia exist within matrices that are entirely equivalent to attached biofilms and are very effective in the biodeterioration of soluble substrates. In the very efficient UASB granules examined by McLeod and his colleagues (8) a central core of methanogenic organisms (Methanotrix) appeared to remove the product (acetate) of a surrounding layer of acetogenic bacteria very efficiently. These acetogens "pulled" propionate and butyrate from essentially heterotrophic organisms in the mantle of the granule. Full development of these complex and very efficient consortia requires three to four months but, once formed, they can degrade organic material and produce methane at truly prodigious rates.

SEDIMENTARY PROCESSES

Biodegradation of organic matter in natural aquatic ecosystems requires the action of consortia of microorganisms in sediments. The upper layers of sediments are oxygenated; the presence of this powerful oxidant and efficient electron acceptor allows the microorganisms to attack even complex substrate molecules. Once the oxygen has been consumed, less effective anaerobic processes must be utilized. The next most favorable electron acceptors are, in order, nitrate, manganese, and ferric iron. These metabolic reactions will also reduce the eH of the sediment sufficiently to allow fermentation reactions to occur. Obviously, at this point the most energetically favorable nutrients will have been consumed as well as many growth factors. Many types of organisms compete with the best growth rate or those which can compete most effectively in limited substrate conditions. The products of fermentation are, as we have seen, dependent on substrate flow and consumption of metabolites. Low energy products such as acetate, CO_2, and hydrogen are produced under optimum conditions. These fermentation reactions are not thermodynamically favorable unless other organisms are present to remove the fermentation products. Inhibition of these consumption reactions result in the production of longer chained fatty acids or alcohols and lactate.

The least favorable electron acceptors still available after fermentation are, in order, sulfate and carbon dioxide. Sulfate reduction is the prevalent form of acetate oxidation in sediments where sufficient sulfate is present. Sulfate reduction also lowers the redox potential in sediments to the point that methanogenesis can proceed. There is evidence that ferrous sulfide masses, produced by precipitation with biogenic hydrogen sulfide, provides a large surface ares and promotes the metabolism of these sedimentary organisms (9). Silt particles also perform this function. The sulfate reducing bacteria are dominant consumers of hydrogen in sediments. These bacteria comprise a heterogeneous grouping of organisms categorized by reduction of sulphur compounds to (usually) H_2S. The sulfate-reducing bacteria utilize a very broad range of substrates and produce several hydrogenases which have a high affinity for hydrogen. Hydrogen tensions are reduced below the level at which other organisms can utilize this energy source. The sulfate-reducing bacteria can also compete successfully with other fermenters in limited substrate conditions and require the cooperation of syntropic hydrogen consumers to grow when sulfate is limiting.

The final degraders of organic matter are the methanogens. These organisms dominate low sulfate anaerobic environments (i.e. < 6-10 mg/l) such as fresh water sediments or lower levels of sediment where sulfate has been depleted. These bacteria utilize acetate, CO_2 and H_2 producing methane. The methanogens often act as syntropic organisms allowing fermentation and acetogenic reactions to proceed. They are fastidious anaerobes and their isolation is extremely difficult.

MICROBIALLY INFLUENCED CORROSION

Early recognition of the phenomenon of microbially influenced corrosion occurred, appropriately enough, in anaerobic soil environments in the Netherlands. Most of the organisms identified as agents of microbially influenced corrosion are anaerobic bacteria originally found in sedimentary environments. The involvement of these bacteria in corrosion reflects the provision of permissive conditions in man-made systems which parallel the natural ecological niches in which these bacteria have evolved.

Many of the operating conditions of industrial equipment are eminently well suited to the development of corrosive bacteria. These systems are often intentionally made anaerobic to reduce the depolarizing effects of oxygen on cathodic processes. Sodium sulfite or ammonium bisulphite are common oxygen scavengers; they provide superior electron acceptors for the sulfate-reducing bacteria. The reduced conditions provide an ideal condition for the development of corrosive species. Even aerated systems frequently develop biofilms in which aerobic bacteria in the upper regions of these adherent communities remove oxygen and provide anaerobic conditions at the metal surface.

The succession of substrates through a mature biofilm parallels the processes which feed bacteria in sediments. Aerobic bacteria degrade complex molecules providing accessible substrates to the anaerobic populations below them. Corrosion inhibitors such as phosphates and nitrates added to some systems can supply vital nutrients to the bacteria. Acetate and other volatile fatty acids are ubiquitous in the brines co-produced with oil and gas.

Creation of anaerobic conditions causes the cathodic reaction mechanism to shift to the production of hydrogen. As we have seen, hydrogen is a vital nutrient in the metabolic processes of sulfate-reducing bacteria. Not coincidentally, the sulfate-reducing bacteria are the most commonly implicated organism in instances of microbially influenced corrosion. The consumption of cathodic hydrogen was identified by early theorists as the mechanism by which corrosion was exacerbated by the sulfate-reducing bacteria. Subsequent conflicting views implicated processes such as the production of H_2S, iron sulfide, or corrosive metabolites such as organic acids or an unidentified phosphorous compound.

These concepts were proposed on the assumption that planctonic bacteria approached the metal surface and attacked it with corrosive ions. Now direct observation has shown us that the bacteria that are present in a corroding system are present as a biofilm (5). This perception immediately suggests that localized microcolonies and consortia of bacteria may develop adjacent surface environments that are local anodes and cathodes. Little (10) has shown that a measurable corrosion potential exists between an uncolonized surface and an adjoining surface colonized by sulfate-reducing bacteria. We suggest that once a corrosion cell is established by chemical or biological action, the development of a heterogeneous biofilm intensifies the corrosion reaction and the concomitant production of hydrogen at the cathode.

Recent work has shown that hydrogenase-positive bacteria can cause severe localized corrosion by depolarizing this cathodic reaction. Cell-free extracts of sulfidogens have been shown to promote the cathodic reactions of hydrogen. Growth of sulfate-reducing bacteria is stimulated by the availability of cathodic hydrogen from corrosion cells and the presence of hydrogenase-positive sulfate-reducing bacteria has been shown to cause the development of a corrosion potential in laboratory studies (Fig. 4).

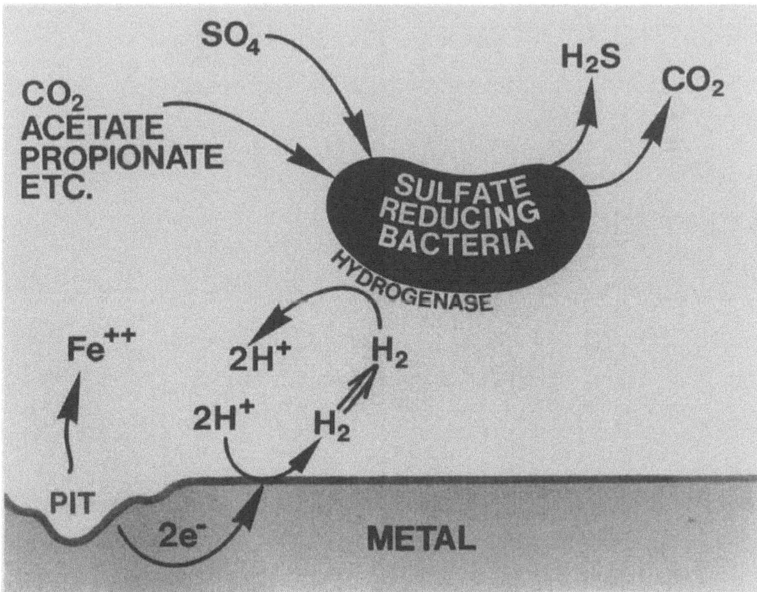

Fig. 4. Corrosion ecology of SRB

The production of hydrogen sulfide by sulfinogenic bacteria produces several conditions which promote corrosion processes. The reduced redox potential increases the electromotive force between cathode and anode. Dissociation of H_2S releases protons for reaction at the cathode. Precipitation of ferrous iron produces a deposit which has an inherent ability to further depolarize cathodic reactions (11). Certain forms of iron sulfide will stimulate hydrogen ion reduction and this will allow the development of ample cathodic hydrogen for consumption by the hydrogenase-positive bacteria. As we have seen, bacteria will colonize ferrous sulfide mats. The close association of bacteria and iron sulfide in the vicinity of the cathodic site produces a deposit with a very large surface area which is electrically connected to the cathode. This results in the development of an extremely large, extremely active cathode. Corrosion engineers are well aware of the corrosion impact of a large cathode coupled to a small anode (Fig. 5). The imbalance in relative areas of anode and cathode stimulates activity at the anode and greatly increases the rate of metal loss in the limited cathode area. The growth of the bacteria-corrosion product matrix encroaches on the anodic area further reducing its area. This results in rapid penetration and early failure of the metal component.

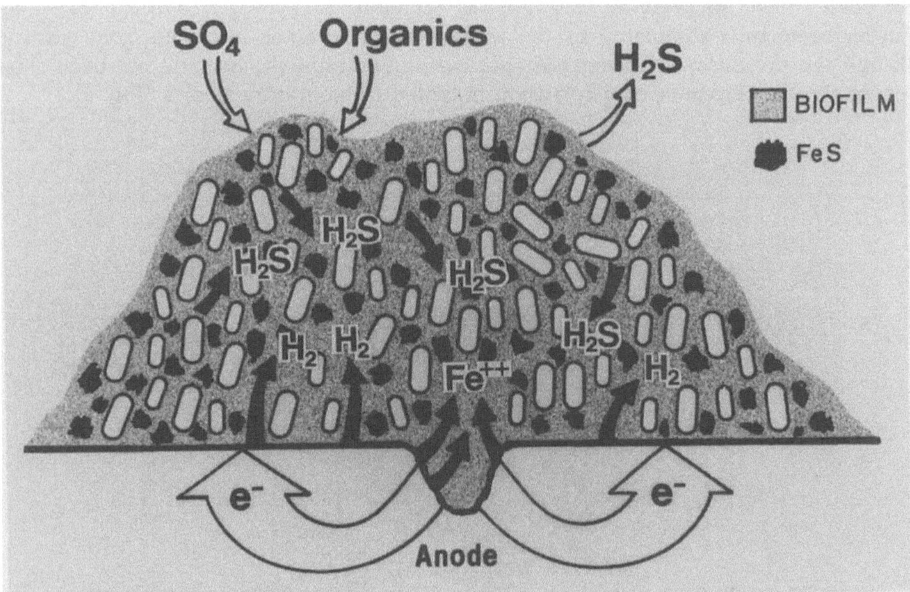

Fig. 5. Cathodic iron sulfide colonized by SRB

The positioning of the hydrogenase-positive organisms in association with the iron sulfide provides them with a ready source of hydrogen. The hydrogenase enzyme has been shown to facilitate the reduction of protons on a metal surface (12). The synergistic effect of this facilitation and the stimulatory influence of iron sulfide on cathodic reactions produce powerful corrosion currents and concomitant hydrogen production. Bacteria deep in a corrosive biofilm are in a nutrient limited condition. The sulfate-reducing bacteria have been shown to function in autotrophic modes under these conditions. Hydrogen can supply all energy requirements and cell carbon can be provided by CO_2 and a small amount of acetate (13). The presence of 10-15 ppm sulfate in the interstitial water of the deposit would be sufficient to allow continued growth of the bacteria.

There is also evidence of microbially influenced corrosion caused by other sedimentary bacteria such as *Shewanella, Clostridium*, and methanogens, which are all avid utilizers of hydrogen. The role of consortia of these bacteria in the anaerobic biodeterioration of metals has not been investigated. The identification of microbially influenced corrosion from such biofilm communities may explain many puzzling instances of corrosion which presently plague industry.

REFERENCES

1   Brown MRW, Williams P (1985) The influence of environment on envelope properties affecting survival of bacteria in infections. Ann. Rev. Microb. 39, 527-556

2   Marshall KC (1976) Interfaces in microbial ecology. Havard Univ. Press, Cambridge

3   Costerton JW, Irvin RT, Cheng KJ (1981) The bacterial glycocalyx in nature and disease. Ann. Rev. Microb. 35, 299-324

4   Morita RJ (1982) Starvation-survival of heterotroph in the marine environment. Adv. Microb. Ecol. 6, 171-198

5   Costerton JW, Cheng KJ, Geesey GG, Ladd TI, Nickel JC, Dasgupta M, Marrie TJ (1987) Bacterial biofilms in nature and disease. Ann Rev. Microb. 41, 435-464

6   Minato H, Suto T (1978) Technique for fractionation of bacteria in rumen microbial ecosystem. II. Attachment of bacteria isolated from bovine rumen to cellulose powder in vitro and elution of attached bacteria therefrom. J. Gen. Appl. Microbiol. 24, 1-16

7   Kudo H, Cheng KJ, Costerton JW (1987) Interactions between *Treponema bryantii* and cellulolytic bacteria in the in vitro degradation of straw cellulose. Can. J. Microbiol. 33, 244-248

8   MacLeod FA, Guiot SR, Costerton JW (1990) Layered structure of bacterial aggregates produced in an upflow anaerobic sludge bed and filter reactor. Appl. Envir. Microbiol. 56, 1298-1307

204

9   Schink B (1988) Principles and limits of anaerobic degradation. In: Zehnder AJB
        (ed) Biology of anaerobic microorganisms. John Wiley, New York; 771-846

10   Gerchakov SM, Little BJ, Wagner P (1986) Probing microbiologically induced
        corrosion. Corrosion 42, 689 - 692

11   Martin RL, Annand RR (1981) Accelerated corrosion of steel by suspended iron
        sulfides in brine. Paper No. 255, Corrosion '81, National Association of Corrosion
        engineers, Toronto

12   Bryant RD, Laishley EJ (1990) The role of hydrogenase in anaerobic corrosion.
        Can. J. Microbiol. 36, 259-264

13   Badziong W, Thauer RK, Zeikus JG (1978) Isolation and characterization of
        Desulfovibrio growing on hydrogen plus sulfate as the sole energy source. Arch.
        Microb. 41, 116-125

INDEX

H. H. Hahn, R. Klute,
University of Karlsruhe

Chemical Water and Wastewater Treatment

Proceedings of the 4th Gothenburg
Symposium 1990
October 1–3, 1990, Madrid, Spain

1990. XI, 560 pp. 335 figs. 110 tabs.
Hardcover DM 178,– ISBN 3-540-53181-5

These proceedings of the 4th Gothenburg
Symposium focus on technology transfer from
chemical treatment theory to practical treatment
of drinking water and industrial or domestic
wastewater. The contributions are devoted to
questions of floc formation and floc separation
as well as problems and practical solutions
associated with chemicals
and dosing control.
Special attention is given to
the combination of chemi-
cal and biological processes
for nutrient removal from
waste water.

W. Fresenius, Taunusstein-Neuhof; **K.-E. Quentin,** Munich;
W. Schneider, Heidenrod (Eds.)

Water Analysis

A Practical Guide to Physico-Chemical, Chemical and Microbiological Water Examination and Quality Assurance

Compiled by W. Schneider

Published by Deutsche Gesellschaft für Technische Zusammenarbeit
(GTZ), Eschborn (FRG)

With contributions by F. J. Bibo, H. Birke, H. Böhm, W. Czysz,
H. Gorbauch, H. J. Hoffmann, H. H. Rump
with Additional Contributions by Other Experts

Translated from the German by A. Gledhill, R. Holland, T. J. Oliver

1988. XXV, 804 pp. 178 figs. 55 tabs. Hardcover DM 160,–
ISBN 3-540-17723-X

The book is written for the practitioner in water analysis and quality
assurance of drinking water. In addition to detailed instructions for sampling and immediate analysis, the book provides a concise presentation
of the theoretical background and of data evaluation. The presented
analytical methods can be applied successfully
with simple equipment as well as in modern
advanced laboratories.
The book is a bench-top laboratory manual
and can be used for instruction in laboratory
staff training programs. It treats the analysis of
organic and inorganic compounds and also
deals with microbiological problems associated
with the guidelines for waste water, surface and
ground water, and drinking water quality.